Transfrontier Conservation in Africa

At the Confluence of Capital, Politics and Nature

'The birth of the Great Limpopo Transfrontier Park today,
tells the citizens of our continent that the AU and NEPAD
are not merely a set of good and grand ideas
whose accomplishment will be in the distant future.
This Transfrontier Park says that
each passing day transforms the dream of
an African Renaissance into reality.'

Thabo Mbeki, 9 December 2002

Transfrontier Conservation in Africa

At the Confluence of Capital, Politics and Nature

MAANO RAMUTSINDELA

MA, PhD

Department of Environmental & Geographical Science
University of Cape Town
South Africa

www.cabi.org

CABI is a trading name of CAB International

CABI Head Office
Nosworthy Way
Wallingford
Oxfordshire OX10 8DE
UK
Tel: +44 (0)1491 832111
Fax: +44 (0)1491 833508
E-mail: info@cabi.org
Website: www.cabi.org

CABI North American Office
745 Atlantic Avenue
8th Floor
Boston, MA 02111
USA

Tel: +1 (617)682-9015

E-mail: cabi-nao@cabi.org

A catalogue record for this book is available from the British Library, London, UK.

A catalogue record for this book is available from the Library of Congress, Washington, DC.

ISBN: 978 1 84593 221 3 (HB)

Maps drawn by Anne Westoby
Project managed by User Friendly
Typeset by Nina Maister, Cape Town, South Africa
Printed and bound by CPI Group (UK) Ltd, Croydon, CR0 4YY

Contents

This book is dedicated to Taki

Acknowledgements

The title of this book has been taken from the honours course, which I developed at the University of Cape Town four years ago. I wish to thank all the students who enrolled for that course for keeping the theme of this book alive and for their engaging questions during the seminars. Over the past years I participated in, and benefited from, various regional workshops and training sessions on transfrontier conservation organised by the Programme for Land and Agrarian Studies (PLAAS), IUCN–ROSA and the Centre for Applied Social Sciences (CASS). The contents of some of the chapters developed from papers presented at the IGU Congress (Glasgow, 2004), the Geography Graduate School Conference (Oulu, 2005), the Annual Conference of the Norwegian Association for Development Research (Äs, 2005) and the ATLAS Africa Conference (Mombasa, 2006). I am grateful to the following individuals for sharing their knowledge and experiences of transfrontier conservation areas with me: Adin Stamelman, Carmel Mbizvo, Chaba Mokuku, David Masilela, Deborah Kahatano, Dominic Mbanjwa, Elna de Beer, Ernest Mokganedi, Evaristo Musonda, Francois Odendaal, Gilberto Vincente, Giuseppe Daconto, Johan Verhoef, Kevan Zunckel, Mark Thornton, Moliehi Shale, Muleso Kharika, Noel de Villiers, Paul Maro, Rabson Dhlodhlo, Refiloe Ntsohi, Sedia Modise and Tankiso Mabote. The book draws on material from my research projects on transfrontier conservation areas, which were funded mainly by the National Research Foundation and supported by funding from the South Africa–Norway Programme and the University of Cape Town, for which I am grateful. To Claire Parfitt I express my gratitude for her encouragement and understanding during the writing of this book.

Preface

As the emergence of transfrontier conservation areas (TFCAs) in southern Africa in the early 1990s enjoyed national and international media coverage, scholars were not ready to engage with the development of such areas since it was a new area of research altogether. The fast pace at which TFCAs were being promoted, together with the lack of analytical tools to engage with them, resulted in much speculative writing. In a rush to explain 'the new phenomenon', such writings ignored the global and regional contexts in which TFCAs were located or to which they were connected. More crucially, there was lack of appreciation of the historical foundations of TFCAs in the southern African region. These weaknesses in scholarship were perpetuated by research that adopted an empiricist analytical gaze that sought to analyse TFCAs in isolation. This book seeks to fill these gaps and to provide new avenues for future research on TFCAs in Africa and beyond. In order to achieve these goals, the book establishes the links between TFCAs in Africa and the global and regional processes from which they flow, and of which they are part; accounts for the use of southern Africa as a launch pad for TFCAs in Africa; and establishes the commonalities of TFCAs in the southern African region.

The book suggests that an adequate analysis of transfrontier conservation initiatives in Africa would have to go beyond explaining specific cases and the circumstances surrounding them if we are to achieve a comprehensive understanding of these initiatives. This is imperative, not least because TFCAs provide the platform on which various interests are pursued at, and between, various scales. The book challenges the emerging and common view of TFCAs as a southern African phenomenon or, at worst, a South African experience. Instead, it argues that TFCAs should be understood in the context of the intersections between the environmental movement and the private sector, and the place and role of the state – individually and collectively – in those intersections. It concludes that events and processes that shaped TFCAs and their emergence in Africa are not perfectly discrete, but are linked to the trajectories of environmentalism and associated practices in a changing socio-political environment.

Maano Ramutsindela

Abbreviations

ARTP	Ais-Ais/Richtersveld Transfrontier Park
AU	African Union
CARTPAN	Central Albertine Rift Transfrontier Protected Area Network
CBD	Convention on Biological Diversity
CBNRM	Community Based Natural Resource Management
CESVI	World Aid from Italy
CI	Conservation International
CITES	Convention on International Trade in Endangered Species of Wild Flora and Fauna
COMESA	Common Market for Eastern and Southern African States
DAP	Drakensberg Approaches Policy
DEAT	Department of Environmental Affairs and Tourism
EAC	East African Community
ECCAS	Economic Community of Central African States
ECOWAS	Economic Community of West African States
ELMS	Environment and Land Management Sector
GEF	Global Environmental Facility
GLTP	Great Limpopo Transfrontier Park
GTZ	German Agency for Technical Co-operation
IGAD	Intergovernmental Authority on Development
IGADD	Intergovernmental Authority on Drought and Development
IGCP	International Gorilla Conservation Programme
IIPT	International Institute for Peace through Tourism
IUCN	World Conservation Union
KAZA	Kavango–Zambezi
KCS	Kalahari Conservation Society
KfW	German Development Bank
KMIA	Kruger Mpumalanga International Airport
KTP	Kgalagadi Transfrontier Park
KTPF	Kalahari Transfrontier Park Foundation
LSDI	Lubombo Spatial Development Initiative
MoU	Memorandum of Understanding
MS	Movement for Survival
NAFTA	North American Free Trade Agreement
NATO	North Atlantic Treaty Organisation
NEPAD	New Partnership for Africa's Development
NGO	Non-Governmental Organisation
PPF	Peace Parks Foundation
RETOSA	Regional Tourism Organisation of Southern Africa
SABSP	Southern African Biodiversity Support Programme
SADC	Southern African Development Community
SADCC	Southern African Development Co-ordination Conference
SANF	Southern African Nature Foundation
SANParks	South African National Parks
SCA	Special Case Area
SDI	Spatial Development Initiative

SNRA	Songimvelo Natural Resource Area
SREAP	Southern Africa Sub-Regional Environmental Action Plan
SWAPO	South West Africa People's Organisation
TBCA	Transboundary Conservation Area
TBNRM	Transboundary Natural Resources Management
TBPA	Transboundary Protected Area
TFCA	Transfrontier Conservation Area
TGLP	Tribal Grazing Land Area
UMA	Arab Maghreb Union
UNEP	United Nations Environment Programme
UNESCO	United Nations Educational, Scientific and Cultural Organisation
USAID	United States Agency for International Development
WILD	International Wilderness Leadership School
WMA	Wildlife Management Area
WWF	World Wide Fund for Nature

Chapter 1

Perspectives on transfrontier conservation areas: an introduction

The development of transfrontier conservation areas[1] (hereafter TFCAs) in Africa in the early 1990s has created much media hype and high interest in academic circles, precisely because of the expectations it has raised[2] and the consequences[3] that might arise from it. The common and, sometimes, overdrawn question being asked from various quarters is whether and how transfrontier conservation initiatives in Africa and elsewhere would contribute to the conservation of biodiversity (Kemp *et al.* 2001; Bennett 2003; Langlois and Ballantine 2005; McKie *et al.* 2005) and the development of impoverished local populations (Katerere *et al.* 2001; Dzingirai 2004; Jones 2005; Van Amerom and Büscher 2005), particularly those living inside or adjacent to areas affected by such initiatives. The question arises mainly as a result of the framing of transfrontier conservation initiatives in the languages of conservation, neo-liberalism and political transformation. Conceptually, the question has its own wider implication: it imposes limits to those seeking to interrogate TFCAs by forcing them to reaffirm the necessity of TFCAs as a point of departure. Put in another way, the question seeks to generate answers about how best transfrontier conservation initiatives should or could work rather than who needs and drives them in the first place. The implication is that the research focus becomes part of the agenda of the environmental movement, with the researcher joining that movement by intention or default.[4] This is so because answers to that question lead to, and fit into, the proliferation of crusades and campaigns for human survival. The campaigns take different forms that are all centred on the notion of sustainable development. On the one hand, proponents of TFCAs view the protection of ecological systems as critical for human survival (Westing 1998). On the other hand, critics of transfrontier conservation initiatives call for ways in which those initiatives could be made more relevant to the livelihoods of, mostly, the rural poor (Jones 2005)[5], and to peace-building efforts (Brock 1991; Ali forthcoming).[6]

An inquiry on what the consequences of transfrontier conservation initiatives are or could be demands, at the minimum, an understanding of what these initiatives are and what drives them. In the context of southern Africa, suggestions have been made that TFCAs – in their various names[7] – emerged from the search for an Afrikaner homeland (Ellis 1994; Draper *et al.* 2004; Hughes 2005), as a mechanism for national identity building in post-apartheid South Africa (Draper *et al.* 2004), an application of the World Conservation Union (IUCN) conservation ideas by the World Wide Fund–South Africa (Simon 2003), an outcome or aspect of globalisation (Duffy 2001, Ramutsindela 2004a), an experiment with selected elements of bioregionalism (Wolmer 2003), or an extension of the community based natural resource management (CBNRM) industry (Dzingirai 2004). While these explanations are sound as far as they go, they raise more serious questions that are yet to be answered. For example, it has been suggested that Afrikaner nationalists embarked on cross-border nature conservation initiatives as a strategy to secure their foothold on land after the demise of the apartheid state. Ellis (1994) argued strongly that the new environmental discourse and the development of a cross-border park on the Mozambique–South Africa–Zimbabwe border were linked up with the broad military strategy of the apartheid state. He concluded that, 'whatever the merits of the new parks in conservation terms, the fact remains that, in the foreseeable future, control of land remains a matter of crucial importance, especially when it is in such a strategic area as the South Africa–Mozambique border' (Ellis 1994, 69). Admittedly, South Africa's military intelligence and Special Forces infiltrated, and even worked through, conservation areas and agencies,[8] but using these factors to explain the rise of the TFCAs in the region is severely limiting for reasons that I advance below.

The link between the development of the Great Limpopo Transfrontier Park (hereafter GLTP) on the Mozambique–South Africa–Zimbabwe border (Figure 1.1) and Afrikaner nationalism has been suggested on the grounds that interest in TFCAs coincided with the search for an Afrikaner homeland, and that the Kruger National Park – which is key to the creation of the GLTP – has historically played a crucial role in the development of Afrikaner nationalism. In this sense, analysts viewed the creation of the GLTP as a strategy for entrenching the Afrikaner's hold on the Kruger National Park in changing political environments. That strategy has been captured by the notion of 'Krugerisation', which is occasionally used outside the official circles. Implicit in 'Krugerisation' is the view that the creation of the GLTP is a process by which political and economic interests that are

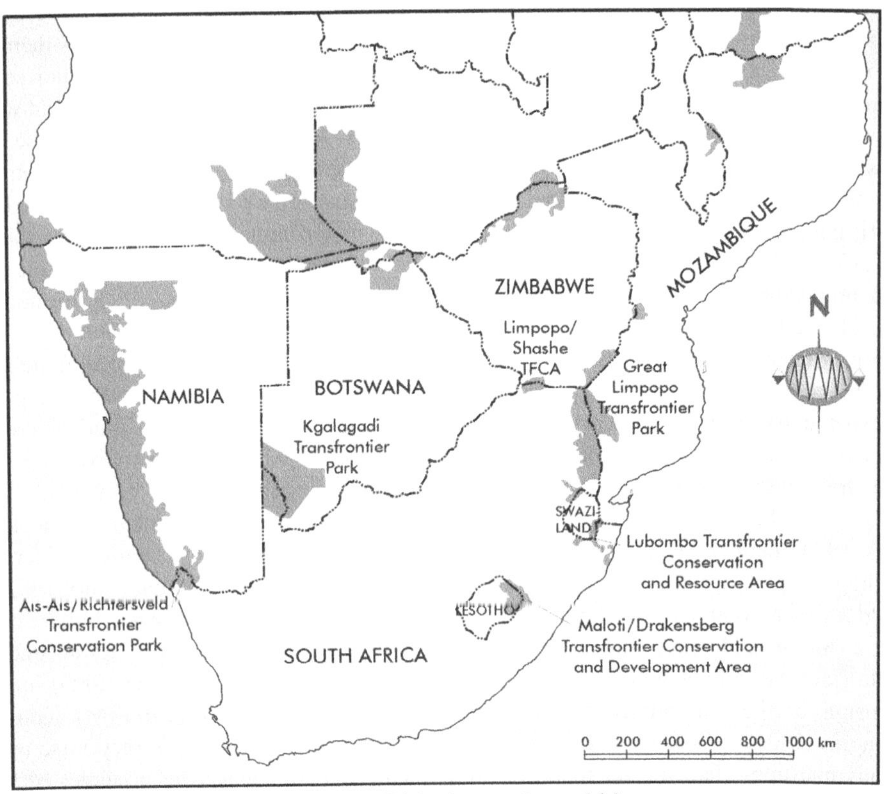

Figure 1.1 TFCAs in southern Africa.

historically embedded in the Kruger National Park are being extended beyond the borders of South Africa, still serving a privileged group at the expense of mostly poor black people. To make the association between the GLTP and Afrikaner nationalism more explicit, analysts have referred to the dominance of the Afrikaners in the Peace Parks Foundation (PPF) (Ellis 1994; Draper *et al.* 2004), which is the main force behind the process of joining contiguous national parks across political boundaries in order to establish transfrontier parks, such as the GLTP.

Admittedly, the interests of the Afrikaner in the GLTP cannot be underestimated. However, the links between Afrikaner nationalism and the GLTP are not as simplistic as they are made to be. My argument here is that while there is merit in the association of Afrikaner nationalism and the Kruger National Park (Carruthers 1995), the relationship between TFCAs and Afrikaner nationalism is less convincing. Afrikaner nationalism has been less homogeneous throughout much of its history, resulting in different Afrikaner-dominated political parties in different time periods (see Greenberg 1987; Giliomee 1992). More importantly, the establishment of TFCAs is not unique to South Africa, and the first official transfrontier park was not even established in Africa.

The pioneers of TFCAs in Africa were involved with conservation projects in southern Africa as early as 1968, and the sites on which TFCAs are established in post-apartheid southern Africa have little or no bearing on the Afrikaners' quest for a homeland. The search for the Afrikaner homeland, the *volkstaat*, was not directed at national parks, but was based mainly on the areas that formed the core of the defunct Afrikaner republics such as the Transvaal (Christopher 1994). Moreover, the main parties that represented the interests of the Afrikaners at the political negotiation table[9] in Kempton Park in 1991–1993, the National Party and the Freedom Front, made proposals for the demarcation of the internal boundaries of the country for a democratic dispensation. The proposals by the two political parties did not suggest that Afrikaner nationalists were preoccupied with national parks, including the Kruger, as part of their envisaged areas of control. Even the more conservative Afrikaners who aimed at establishing their own *volkstaat* under the banner of the constitutionally sanctioned *Volkstaat* Council (see Ramutsindela 1998) suggested maps in which the TFCAs were not central (Figure 1.2). In no way do I suggest that Afrikaners were less interested in national parks during the transition to a democratic South Africa. Rather, the point is that TFCAs were not central to the geographical expression of Afrikaner interests at the time. Even if this were the case with regard to the Kruger National Park, how would we relate Afrikaner nationalism to other TFCAs such as the Maloti/Drakensberg and the Kgalagadi. Moreover, the Afrikaner-dominated PPF is not the only organisation involved in the creation of TFCAs in the sub-region. Other bodies and institutions such as the World Wide Fund for Nature (WWF), Conservation International (CI), the Global Environmental Facility (GEF) and the like, are active players in the creation of TFCAs in southern Africa and beyond.

It could be argued that interests in cross-border nature conservation projects in southern Africa emerged long before the current idea of an Afrikaner homeland was firmly established. In the 1920s, General Jan Smuts was interested in establishing a wildlife sanctuary in Mapungubwe,[10] and in linking that sanctuary with conservation areas in neighbouring countries (Tiley 2004). Smuts was also interested in a cross-border nature conservation project that would expand the Kruger National Park beyond the borders of South Africa. It is said that the Mozambican ecologist, Gomes, was interested in joining Mozambique's protected areas with the Kruger National Park to establish some form of a cross-border nature conservation area in the 1930s (South African National Parks 1997).

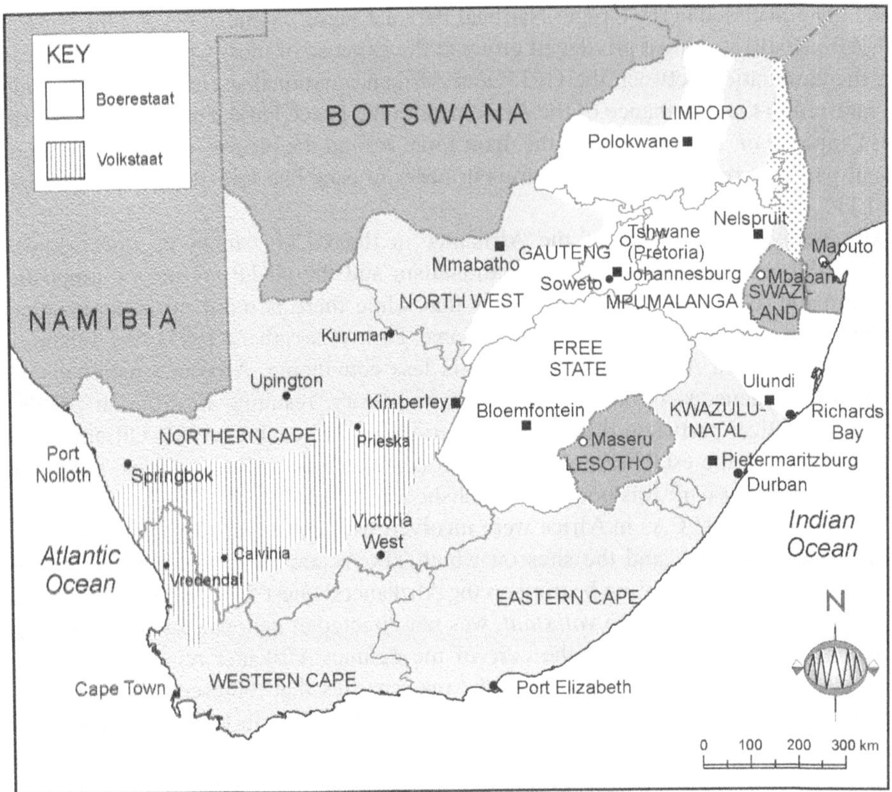

Figure 1.2 Proposed Afrikaner 'homelands'.

An adequate analysis of transfrontier conservation initiatives in Africa would have to go beyond explaining specific cases and the circumstances surrounding them if we are to achieve a comprehensive understanding of these initiatives. The first step towards achieving this goal is to discard the view of TFCAs as a southern African phenomenon or, at worst, a South African experience. Failure to do this will most likely insulate transfrontier conservation initiatives in the region from their wider ideological milieus, as is currently the case with most studies. TFCAs should be understood in the context of intersections between the environmental movement[11] and the private sector, and the place and role of the state in those intersections. In other words, there is a need to establish the roots of TFCAs within and beyond the borders of the African continent. This will certainly help us to develop meaningful analytical frameworks for, and approaches to, TFCAs in Africa.

It would also be useful to understand the commonalities and differences among transfrontier conservation initiatives, if we are to avoid producing insufficient general conclusions. For instance, most explanations of the emergence of TFCAs in southern Africa are often generalised from the experiences of the GLTP, mainly because of its popularity and the extensive resource base that is accessible to researchers compared to most other TFCAs in the region. The GLTP is increasingly being used to generalise TFCAs in the region despite the evidence that the first officially sanctioned example of a TFCA in the region and the continent was the Kgalagadi Transfrontier Park, which was formally established in May 2000. More crucially, most analyses are limited to the post-1990 southern African situation, with casual references to the colonists' attempts at cross-border nature conservation schemes (South African National Parks 1997; Wolmer 2003). A closer inspection of most of the sites on which transfrontier conservation initiatives received their official seal in post-apartheid southern Africa reveals that bioregional planning of some sort took place long before the official launching of TFCAs in the region. Generally, African states were involved in managing some of their shared natural resources before the creation of TFCAs. That is to say that transfrontier conservation in Africa has historical roots that have been completely ignored, leading to the under-theorisation of either the processes towards the establishment of contemporary TFCAs or the ideological foundations on which they are anchored.

A more helpful approach towards understanding TFCAs in the region and the African continent would be to trace their development in the region, particularly different activities that preceded the formalisation of cross-border nature conservation projects in the early 1990s. Identifying and explaining those activities is important, because TFCAs were not established in a vacuum, but were systematically developed from existing nature conservation projects, including national parks. Questions of academic interest are whether different transfrontier conservation initiatives are in fact informed by the same ideas, and what similarities and differences do TFCAs have in different parts of Africa and the world at large. In seeking answers to these pertinent questions we might arrive at a deeper understanding of the rise and operation of TFCAs in Africa. I do not suggest for a moment that there is a single explanation for the emergence of TFCAs in the continent nor that one perspective is more important than the others. Neither do I privilege the use of any particular theory above others. Rather, I suggest that the analyses of, and perspectives on, TFCAs in Africa that have been offered in the first decade of TFCAs in the continent are insufficient.

My contention is that events and processes that shaped TFCAs and their emergence in Africa are not perfectly discrete, but are linked to the trajectories of environmentalism and associated practices in a changing socio-political environment. Arguably, TFCAs in Africa are by and large a product of ideas about ecological integrity while, at the same time, representing ways by which protected areas can be created and expanded through the merging of conservation and development ideals with the appropriate political and financial backing.

The aim of the book

The current but nascent perspectives on TFCAs in Africa suffer from four main weaknesses, the first being the failure to acknowledge the connections between TFCAs in the continent and the external processes from which they flow, and of which they are part. Secondly, almost all analysts of TFCAs in southern Africa ignore the roles that the frontrunner of the PPF, the Southern African Nature Foundation, played in laying the foundation for contemporary TFCAs in the region. Thirdly, there is lack of information on why and how southern Africa stood out as the launching pad for TFCAs in the continent. Fourthly, there is no systematic account for why TFCAs have been embraced by different interest groups. This book seeks to make a contribution towards addressing these weaknesses by first providing a background to different ideologies that are closely related to the TFCAs. In particular, the book sketches out the agendas and strategies of the environmental movement and discusses the relevance of environmental security, regionalism and ecotourism to the analysis of TFCAs.

The principal objectives of the book are to provide background to the ideologies and practices underpinning the content and form of TFCAs in Africa, to historicise the PPF

– as the main actor in TFCAs – and its global linkages, to analyse regional dynamics and their relationships with the rise of TFCAs, and to account for the apparent lack of resistance towards the creation of TFCAs in southern Africa. Though the book covers different parts of Africa, most of the examples of TFCAs are drawn from southern Africa, simply because the region has been a launching pad for TFCAs in the continent. The book treats the relatively slow pace of TFCAs in other politico-economic regions of the continent as part of the explanation of the forces and motives behind TFCAs in Africa. In other words, the geographical difference in the spread of TFCAs in Africa is part of the explanation of why and how TFCAs have emerged in the continent.

Organisation of the book

Rather than giving detailed accounts of each TFCA in Africa, the approach taken in this book is to generate overarching ideas and trends from the empirical evidence from the author's own research, and from various studies and unpublished material. This approach has been taken in an attempt to move research on TFCAs beyond the constraints of empiricism while, at the same time, acknowledging the value of empirical evidence in the formulation of theory. In line with this approach, each chapter focuses on a theme that goes beyond specific TFCA sites. The aim is to offer what Castree (2005, 544) calls a 'concretised abstract discussion'.

Chapter 2 On the ideological foundations of transfrontier conservation areas

The statement by the former South African President, Nelson Mandela (2001) in which he said, 'I know of no political movement, no philosophy, no ideology, which does not agree with the peace parks concept as we see it going into fruition today',[12] has been used by proponents of TFCAs to validate the ideas behind the creation of TFCAs, particularly in Africa. This was possible because the quotation is strategically silent about which philosophy actually does support TFCAs. To expose the philosophy behind TFCAs would make TFCAs vulnerable to the weaknesses of that philosophy. It may also be that if a particular philosophy is overtly associated with TFCAs, the tenets of that philosophy might be challenged in places where TFCAs are to be established. The task of scholars of TFCAs is therefore to identify and engage with ideologies that underpin or are relevant to the analysis of TFCAs. The focus of this chapter is on the environmental movement and the general foundation it has laid for the development of ecoregions of different kinds. The relevance of the environmental movement to the discussion on TFCAs is that the notions of environmental crisis, together with the responses to that crisis by the environmental movement and its allied institutions, are central to the mobilisation of ideas that underpin the creation of TFCAs. The premise of this chapter is that a careful analysis of the environmental movement is helpful for understanding the rationales for, and interests in, TFCAs. The chapter suggests that TFCAs are closely connected to the notion of the environmental crisis and the global solutions that have emerged from the interpretation and framing of that crisis.

The environmental movement responded to the environmental crisis in different ways. The three main responses that are relevant to our discussion are, firstly, the development of non-governmental organisations (NGOs) to promote ecological integrity from outside the state. Secondly, the infiltration of the state by political parties that are founded on ecological grounds, the 'green parties'. The ultimate aim of the green parties is to create a 'green state'. Thirdly, a complete restructuring of the nation-state in favour of the development of grass-roots autonomous regions. For Africa, interpretations of the environmental crisis are directly linked with the poor socio-political conditions in that continent. The absence of 'green parties' and the movement for grass-roots autonomous regions (bioregional movement) means that environmental NGOs, such as the PPF, are the main actors in the creation of TFCAs.

Chapter 3 Peace through ecology? A soft approach to hard realities

The concept of peace parks has been used in southern Africa to highlight the significance of TFCAs beyond the imperatives of ecological integrity. Notably, peace parks are not an African phenomenon as they have been established in different parts of the world. The Waterton/Glacier International Peace Park on the US–Canada border is generally considered the forerunner of, and model for, peace parks globally. In this chapter I suggest that TFCAs in Africa should be understood in the context of global models of peace parks. Two main types of peace parks stand out: the first being parks that are established within particular countries as commemorative sites or as a way of preserving particular local sites; the second involves the amalgamation of contiguous national parks across state borders, using the need to preserve the ecological integrity of areas as a guiding principle. The discussion in this chapter and the volume as whole focuses exclusively on the second category of peace parks.[13] In order to emphasise the application of the cross-border peace park model to the African context, the chapter highlights three main general propositions of peace parks. These are that: (i) international boundaries are more important for peace parks than sub-national ones; (ii) boundaries around peace parks are natural; and (iii) the environment is indissolubly interlinked with instability and peace.

The end of the Cold War heralded shifts in the concept of national security, a process that was reinforced in southern Africa by the collapse of the apartheid state. New ideas about security incorporated the environment as an element of international security. For Africa, it is assumed that environmental problems will contribute to the insecurity of the continent, hence there is a view that Africa's strategic vision should include regional and subregional co-operation on shared problems. TFCAs are therefore seen as part of the strategy to bring about the much-needed peace on the continent by acting as symbols of peace and a platform for inter-state co-operation. The chapter concludes that, in Africa, proponents of peace parks have used boundary narratives, which are relevant to ongoing discussions about the future of Africa and the unity of its people and their environment to promote TFCAs as politically, economically, socially and ecologically relevant to the continent.

Chapter 4 Local initiatives and their regional and global connections

Proceeding from the peace park model and its adoption in Africa, this chapter sketches the development of contemporary TFCAs from disparate nature conservation projects in southern Africa. The focus of the chapter is twofold. The first part focuses on the background and activities of the Southern African Nature Foundation as a forerunner of the PPF. That background is important for understanding the development of TFCAs in the region and the rest of the continent on the one hand, and to establish the links between TFCAs in Africa and the rest of the world, on the other hand. The chapter uses the background of the Southern African Nature Foundation to argue that the map of TFCAs in southern Africa has its roots in activities that occurred long before the formalisation of TFCAs, and that the links between TFCAs and the private sector have a long pedigree that studies on TFCAs in Africa should acknowledge.

The second part of the chapter focuses on the rise of transfrontier parks in southern Africa. It captures the development of the first transfrontier park, the Kgalagadi through to the proposed Limpopo/Shashe,[14] to suggest that transfrontier parks in southern Africa reinforce the dominance of national parks in cross-border nature conservation. The Kgalagadi Transfrontier Park therefore provided the first model for TFCAs[15] in which national parks are pillars of an ecoregion. The model also suggests a shift from TFCAs to transfrontier parks, a process that has led to the establishment of national parks in areas earmarked for TFCAs. The consequence of that shift is the move towards the creation of ecoregions that are devoid of human settlements. Where such human settlement exists, the plan is to remove people from the areas that are designated for transfrontier parks, as I show in Chapter 6.

Chapter 5 Southern African Development Community: a regional catalyst

The launching of TFCAs in southern Africa and the delays in the establishment of TFCAs in other parts of Africa warrant some explanation. That explanation is important, not least because there is a well-established but ill-founded tendency to offer generalised solutions to Africa's varied and context-specific problems, including those related to the continent's natural environment. The question that arises from the establishment of TFCAs in southern Africa, and the delay in creating them in the continent's other regions, is what is unique about southern Africa. This chapter engages with that question by referring to ecological conditions that are optimal for the establishment of TFCAs in other parts of Africa, particularly, East and West Africa. The chapter notes that the Convention on Biological Diversity (CBD) that was adopted at the Earth Summit in Rio in 1992 sanctioned the merging of conservation and development that became a benchmark for developing national environmental policies. A review of the status of signature, ratification, accession, acceptance and approval of the CBD in Africa reveals that southern African states, with the exception of the island states of Mauritius and Seychelles, did not necessarily take the lead in ratifying the CBD. Stated differently, most countries in other regions of Africa ratified the CBD before southern African states. The same can be said about the process of regionalism. Eastern and western African states embarked on the process of economic and political regionalism long before the Southern African Development Community (SADC) was established in 1992. Nevertheless, regionalism in East and West Africa did not strongly support or facilitate the development of TFCAs in those regions, despite the fact that environmental and socio-economic conditions favoured the establishment of TFCAs in those regions. Admittedly, regionalism is a differentiated process in the continent hence the outcomes of that process are bound to be different. However, most regions in Africa have been involved in one form of cross-border natural resources management initiative or another,[16] yet those initiatives did not necessarily translate into TFCAs.

The chapter suggests that regionalism in southern Africa opened up opportunities for creating TFCAs and for inserting them into the regional environmental agenda. The Washington-based Biodiversity Support Programme commissioned a study on the development of transboundary natural resource management (TBNRM) areas in southern Africa, which identified 17 potential terrestrial TBNRM areas. Most of those areas had already been proposed by the PPF. The SADC structures became instrumental in formalising the TFCAs plans that had been developed outside those structures.

Chapter 6 Silencing community struggles

Despite the history of community struggles for natural resource rights in southern Africa, there has not been any visible protest[17] against the establishment of TFCAs in the region. This is rather surprising, because TFCAs involve some of the national parks that were – and some still are – being contested by local communities. Moreover, some of the TFCAs directly affect the communities living inside and adjacent to areas designated for TFCAs by, say, resettlement. Against this backdrop, the purpose of this chapter is to account for the lack of community protests against the TFCAs. The chapter seeks to offer a theoretical explanation that is grounded on the practices of proponents of TFCAs. Theoretically, the chapter explains the lack of community protests by drawing on insights from a well-established body of work on scale analysis, the aim being to locate the community within the construction of the scale at which TFCAs are established. In other words, we need to observe processes that underscore the creation of TFCAs at the appropriate level so as to understand how those processes open or close the spaces of community engagement. The chapter refers to construction of scales to make three related arguments. Firstly, it argues that TFCAs manifest a particular construction of scales that disengages local communities, hence those communities are unable to contest or engage with the TFCA process. That is to say that the local communities are marginal to the processes that produce TFCAs. Secondly, it argues that TFCAs involve the double rearticulation of political scales. Whereas the need for the joint protection and management of biodiversity and fund raising forces the state to scale up its regulatory functions to a regional entity, the local demands and struggles for natural resources result in the scaling down of the functions of the state. Individual states remain responsible for their citizens in terms of their constitutions. Apart from signing the necessary legislative documents, states play a crucial role in realigning their policies to the ideals of transfrontier conservation. The third argument is that at the sub-national level, community-based natural resources management projects mediate conflict over natural resources while, at the same time, ensuring that communal land functions as corridors for transfrontier conservation.

To substantiate these arguments, the chapter draws on the experiences in Botswana, Mozambique, Namibia, South Africa and Zimbabwe. Specifically, Wildlife Management Areas in Botswana are referred to in order to demonstrate the ways in which state policies are appropriated to support the creation of TFCAs. Namibia's conservancies are useful for understanding the realignment of land use options for the promotion of TFCAs, while the drive for privatisation and ecotourism in South Africa highlights the trend towards developing national policies that reflect interests in TFCAs. Central to the discussion of the experiences from those countries is the claim that the TFCAs involve a process of scaling that disengages local communities who are supposed to be the beneficiaries of TFCAs.

Chapter 7 The renaissance of the bush: the reinvention of the Cape to Cairo route in contemporary Africa

The main focus of this chapter is on the links between TFCAs and ecotourism in Africa. Those links are established by, firstly, analysing one of the famous tourism routes in Africa, the Cape to Cairo route. The analysis begins by considering the idea of the Cape to Cairo in its colonial context, the purpose being to excavate the content of that idea in the age of imperialism so as to reassess its re-emergence in 21st century Africa. It has been suggested that Cecil Rhodes embraced the idea of the Cape to Cairo, because it expressed all he had ever dreamt of, namely, the establishment of a British World Federation in which a British Pax Africana formed an important part. The Cape to Cairo idea was initially inspired by commercial interests, such as mining, before it developed into a sharp colonising instrument. Another commercial interest that is significant for the theme of this chapter is overland tourism. Contemporary TFCAs promote ecotourism along the Cape to Cairo axis by establishing an unbroken zone of protected areas from the Cape to the north of the continent. Other routes have been planned to promote tourism in TFCAs.

The chapter concludes that the reappearance of the Cape to Cairo idea as a nature conservation and tourism scheme begs the question of the manner in, and extent to, which colonial ideas are reinvented. The chapter suggests that the construction of the Cape to Cairo idea in different historical moments provides a useful framework for analysing the colonial presence. It argues that the British imperialist vision of connecting Africa through the railway and the telegraph, and contemporary trans-Africa eco-based conservation schemes both have common political and economic resonance, and deploy similar tactics. The fact that southern Africa has been a launching pad for ideas for a continent-wide linkage in different historical moments suggests complex ways in which geography is embroiled in the propagation of visions for Africa.

Chapter 8 Mapping Africa's future

O'Riordan (1996, 458) correctly pointed out that 'environmentalism thrives by being injected with ideas and adherents of other social movements, whose aims and political purpose overlap with environmentally more stable and socially more peaceful and just futures'. In essence, environmentalism provides a platform on which possible futures are forged on environmental grounds. For ecologists and like-minded individuals and institutions, the future of humankind hinges on the proper working of ecosystems. The Gaian perspective of nature has even gone further to suggest that humans have to find their place in the natural order that existed long before it was interrupted by humans. Within the social sciences, human survival is conceptualised in terms of socio-political needs and aspirations, thereby dismissing a narrow biophysical view of the environment. Against this backdrop, TFCAs embrace different layers of environmentalism,[18] hence they lead to a fusion of interests among ecologists, politicians, social movements, capitalists and local communities. That fusion is facilitated by the diverse roles that ecology plays as an 'institutional shaman that can be induced to pronounce natural whatever we wish to espouse' (Everden, cited in Pedynowski 2003, 807). In Africa, the coalescence of interests in TFCAs has moved from local projects to continental schemes such as the New Partnership for Africa's Development (NEPAD). Clearly, the intention is to define the future of Africa's natural resources and to suggest how and by whom those resources would be controlled and used across state borders.

Chapter 9 Postscript: transfrontier parks and the legacy of Rupert

Comments that were made on the death of Anton Rupert in January 2006 do not only reveal that Rupert was one of the main founders of TFCAs in southern Africa, but also suggest that his vision of the African landscape has been embraced by politicians, the private sector and NGOs. The elevation of Rupert to the status of a hero implies that a critical assessment of TFCAs in Africa on ideological grounds is a challenging task.

Endnotes

[1] In this volume, a transfrontier conservation area is regarded as a conservation area made up of protected areas and other land use types across international boundaries.

[2] These include a tourism boom (*City Press* 2005), better management of protected areas and peace among participating states.

[3] Concerns have been raised that TFCAs might lead to the marginalisation of local residents, loss of land rights, the dominance of one country over others, and so on.

[4] The 'moral imperative' of TFCAs accounts largely for the absence of an intellectual project that seeks to argue against the necessity of TFCAs in southern Africa.

[5] Jones (2005) laments that there has been minimal research on the impact of TFCAs on local communities, and used the case of Mbangweni community on the South African side of the Maloti/Drakensberg transfrontier conservation area to demonstrate that local communities could experience decreased access to social, natural and economic resources as a result of the TFCA initiative.

[6] Within the social sciences, there is a nascent but growing body of work that seeks to critique the socio-political objectives of transfrontier conservation initiatives (see Duffy 2001; Simon 2003; Wolmer 2003; Ramutsindela 2004a; Van Ameron and Büscher 2005).

[7] Some of the common names used in southern Africa are peace parks and transboundary natural resources management areas.

[8] It has been reported that Colonel Gert Otto of the apartheid-era South African Defence Force attempted to promote a cross-border park between the Kruger National Park and part of Mozambique in order to facilitate the flow of intelligence material to the South African security forces (South African National Parks 1997).

[9] Following the unbanning of the liberation movements in South Africa, political parties and civil society were involved in negotiating the new political dispensation. The main parties at the negotiation table were the then ruling National Party and the African National Congress (see Sparks 2003).

[10] Tiley (2004) has described Jan Smuts as Mapungubwe's visionary.

[11] In its various forms.

[12] Referring to the release of elephants from South Africa's Kruger National Park into Mozambique to mark the first step towards the creation of a borderless landscape of the GLTP.

[13] For reasons that are discussed in full in the chapter.

[14] Transfrontier Park.

[15] In the southern African region

[16] This is particularly true in the water sector.

[17] In no way do I suggest that all local communities are happy with the establishment of TFCAs that impact on their lives. Rather, my view is that no form of community dissatisfaction has translated into a campaign against the establishment of a particular TFCA.

[18] According to O'Riordan (1996) the four layers of environmentalism are underpinned by views of nature, intervention, political economy, and the fusion of social movement.

Chapter 2

On the ideological foundations of transfrontier conservation areas

The multi-layered bases on which transfrontier conservation areas are founded and promoted challenge any simplistic explanation of TFCAs. The layers range from interests in promoting ecological integrity and governance through international relations and peace to socio-economic development. A full explanation of TFCAs should therefore deal with the ideas on which they are founded and the various actors that are involved in, and with, them. The focus of this chapter is on the environmental movement and the general foundation it has laid for the development of ecoregions. The relevance of the environmental movement to the discussion on TFCAs is that the notions of environmental crisis, together with the responses to that crisis by the environmental movement and allied institutions, are central to the mobilisation of ideas that underpin the creation of TFCAs. In addition, those notions and responses largely account for the coalescence of interests among disparate groups and institutions. The premise of this chapter is that a careful analysis of the environmental movement is helpful for understanding the rationales for, and interests in, TFCAs.

Environmental crisis and global solutions

The former President of the World Bank, James Wolfensohn, and his colleagues from Conservation International and the Global Environmental Facility, issued a warning that 'we are failing to stem the lethal dynamic of chronic poverty and growing population, which is destroying species a thousand times faster than ever before' (Wolfensohn *et al.* 2000, 38). Indeed, studies demonstrating the catastrophic extinction of animal and plant species are common in the natural sciences, and any contrary view to the environmental crisis[1] is often dismissed as ill informed or speculative (FitzRoy and Smith 2004). The criticism levelled against, and the dismissal of, Lomborg's *Skeptical Environmentalist* (2001), is an obvious example of some such reactions. Critiquing the notion of environmental crisis is misconstrued as a deliberate ignorance of scientific truths or a delay in solving a life-threatening crisis. For environmentalism there is no time to waste, as global ecosystems require millions of years to recover and therefore no further destruction of ecosystems can be allowed to continue. The crisis is not only about the destruction of fauna and flora, but is also about the disintegration of ecological systems that are a source of all life. In support of that perspective, Brunckhorst (2002, 1) remarked that, 'our future, along with the rest of life on earth, depends on ... the services provided by fully functional ecological systems'. In this way, the environmental crisis translates into a threat to life, thereby bringing together diverse groups of people who are concerned with those 'threats': anti-war, anti-nuclear proliferation, anti-industrialisation, and the list is endless. The point here is that the environmental movement has been a broad church since the elevation of local environmental issues to the public domain. The *New Republic* aptly described the American ecological movement in 1970 as 'the biggest assortment of ill-matched allies since the crusades – young and old, radicals of the left and right, liberals and conservatives, humanists and scientists, atheists and deists' (cited in McCormick 1989, x). The 'assortment of ill-matched allies' is one of the defining characteristics of the environmental movement. As I will show later, the same can be said about transfrontier conservation initiatives, because the initiatives pursue the logic of the environmental movement. For the moment I focus on the nature and agendas of the environmental movement in order to understand how and why the environmental movement brings people of diverse backgrounds together.

Proceeding from the view that the environmental movement has no clear beginning, McCormick (1989, 37) regards the United Nations Scientific Conference on the Conservation and Utilisation of Resources (UNCCUR) that was held at Lake Success (New York) between 17 August and 6 September 1949 as 'the first major landmark in the rise of the international environmental movement'. The theme of the conference centred on the adequacy of natural resources to meet growing demands – a rehearsal of Thomas Malthus' doctrine of the inverse relationships between population growth and food supply. Though Malthus' population-resource doctrine has been criticised for being too simplistic and ignoring technological advancement in food production, his doctrine provided the solid foundation on which ideas about the environmental crisis were developed and nurtured. It is therefore not surprising that population issues have since become the first point of reference in the discussion of environmental crises and proposed solutions. The Club of Rome – an informal organisation that originated from the meeting[2] of 30 individuals in the *Academia dei Lincei* in Rome in April 1968 – added dimensions to the environmental crisis when it sought to understand the interdependence of economic, political, natural and social components of the world system in order to promote new policy initiatives and actions (Meadows *et al.* 1972). The Club's project[3] on the predicament of mankind sought to 'examine the complex of problems troubling men of all nations: poverty in the midst of plenty; degradation of the environment; loss of faith in institutions; uncontrolled urban spread; insecurity of employment; alienation of youth; rejection of traditional values; and inflation and other monetary and economic disruptions' (Meadows *et al.* 1972, 10). It used population growth and resource exploitation to develop models, which suggested the need for an orderly transition to a global equilibrium between population growth and resource usage (Meadows *et al.* 1972). It was thought that such a state of equilibrium would release people from the pressures for survival, and would ensure that people direct their energy and ingenuity towards solving other problems. That idea ties well with Maslow's hierarchy of needs.

The manner in which the ecological problems were to be solved, with respect to human survival, has been, and still remains a source of contention. In Britain a group of scientists endorsed the views of the Club of Rome and sought to offer a 'Blueprint for Survival', which was published as a special issue of the *Ecologist* in January 1972 under the editorship of Edward Goldsmith. The publication heralded the formation of the Movement for Survival (MS), which sought to champion a new age in which humans 'will learn to live with the rest of nature rather than against it' (*Ecologist* 1972, 1). Its point of departure was the defect of industrialism, as the following quotation suggests:

> the principal defect of the industrial way of life with its ethos of expansion is that it is not sustainable … We can be certain, however, that sooner or later it will end and that it will do so in one or two ways: either against our will, in a succession of

> famines, epidemics, social crises and wars; or because we want to – because we wish to create a society which will not impose hardships and cruelty upon our children – in a succession of thoughtful, human and measured changes (*Ecologist* 1972, 2).

It was hoped that the MS would become an international movement with a new philosophy of life that would provide the basis for four principal conditions towards a stable society: (i) minimum disruption of ecological processes, (ii) maximum conservation of materials and energy, (iii) stabilising the population, and (iv) establishing a decentralised system of government. According to the Blueprint for Survival, ecological processes were being disrupted by 'introducing into them either substances that are foreign to them or the correct ones in the wrong quantities' (*Ecologist* 1972, 9). The solution to this included restrictions on the use of pesticides and fertilisers. Of significance to the theme of this chapter is the Blueprint for Survival's injunction that wilderness areas and wetlands should be conserved at all costs. It viewed areas of wilderness as the world centres of diversity – the gene banks. It also proposed the creation of a new system of government based on decentralisation, the emphasis being on the creation of small communities that will guarantee both the minimal impact of humans on the environment and community participation. The Blueprint for Survival therefore suggested the restructuring of society and governance on ecological grounds, presenting decentralisation as an authoritative organisational structure of society for the attainment of 'global equilibrium'. How these objectives were to be achieved generated different responses. The three main responses that are relevant for our discussion are, firstly, the development of non-governmental organisations (NGOs) to promote ecological integrity from outside the state. Secondly, the infiltration of the state by political parties that are founded on ecological grounds, the 'green parties' with the aim for creating a 'green state'. Thirdly, a complete restructuring of the nation-state in favour of the development of grass-roots autonomous regions. Accordingly, Dryzek *et al.* (2003) have categorised environmental groups according to those that are unconcerned with the state, those that co-operated with the state and those initially based in oppositional civil society. These categories imply that the environmental movement does not fit nicely into life-cycle progression theory in which it is assumed that social movements begin from oppositional and confrontational positions and later on gravitate towards the state. Instead, the shape and fortunes of the environmental movement are influenced by the structural response of the state and the state's general orientation to societal needs (Kitschelt 1986; Dryzek *et al.* 2003).

Promoting ecological integrity: acting from outside

It should be noted that the attempt to develop the MS was based on the view that there were already environmental interest groups in existence in different parts of the world, which would form the building blocks for the international movement. According to McCormick (1989), the first environmental interest group – the Commons, Footpaths and Open Spaces Preservation Society – was founded in Britain in 1865. Since then, environmental interest groups have mushroomed all over the world. Of importance to our discussion in this chapter is the point that most of these groups opted to pursue an ecological agenda from outside government structures. For instance, at the height of post-world war reconstruction, the conservation network in continental Europe, the International Office for the Protection of Nature (IOPN), successfully opposed Britain and America's attempts to establish an international nature protection organisation under the auspices of the United Nations system. The IOPN opted to pursue the environmental agenda outside state systems. In reality such an agenda can only be pursued through the governments, leading to the hybrid nature of environmental interest groups, as has been the case with the constitution of the Provisional International Union for the Protection of Nature (IUPN) that was signed on 5 October 1948. As such, the purpose of the IUPN was to facilitate co-operation between governments and national and international organisations. It should be noted that the IUPN is a product of the Swiss-sponsored meeting that was attended by 70 delegates from 24 countries between 28 June and 3 July 1947 in Brunnen. The organisation changed its name to the International Union for the Conservation of Nature and Natural Resources (IUCN) in 1956 (McCormick 1989). The point here is that those environmental organisations that opted to operate outside the state were caught in the unenviable position of having to co-operate with the same governments that were, in the organisations' view, destroying – directly or indirectly – the ecosystems on which human survival depends. Environmental organisations deploying conventional lobbying strategies became increasingly less oppositional to the state, as they sought to access bureaucratic and informal opportunities offered by government officials. This is the case with moderate mainstream environmental organisations such as the Wilderness Society (United States), Norwegian Society for the Conservation of Nature (Norway), the Ramblers Association (United Kingdom), and the German League for Nature Conservation and Environmental Protection (Germany) (Dryzek *et al.* 2003).

It would be unrealistic to assume that the influence over the environmental movement is one-directional. Over the years, environmental NGOs have operated in partnership with the private sector to exert influence and pressure on the state. For example, the IUCN has used public platforms such as its World Parks Congresses to suggest directions that states should follow in matters of nature conservation. Those directions have found their way into state policies and strategies. A clear example of this is the IUCN's recommendation that each country should set aside the minimum of 10% of its land for protected areas. In government circles, the recommendation has become a necessity rather than a suggestion by a non-state actor. The IUCN considers itself as the world's largest[4] and most important conservation network that seeks to influence, encourage and assist societies in conserving biodiversity. The link between environmental NGOs and the private sector means that governments can access funds from donors if they follow the guidelines of environmental NGOs. This is so because most funding[5] for environmental matters comes through those NGOs.

Operating within and with the state

The second response towards the environmental crisis was to follow the Blueprint's suggestion that the MS should be committed to act at a national level, including contesting elections. At the time of drawing up the Blueprint, there were no clear models for the kinds of states that Green parties were to develop and administer, and how the aim for a green state could be achieved. To give substance to the feasibility and viability of such models, the *Ecologist* published President Nyerere's model of a state based on the vision of a self-reliant independent Tanzania. In writing for the *Ecologist*, Jimoh Omo-Fadaka (1972, 7) painted a picture of the Tanzanian state that appealed to the authors of the Blueprint by suggesting that:

> Tanzania does not want to turn into a western society if this means the constant pursuit of materialist goals and the social divisions produced by the differential accumulation of wealth. It does not want to turn into a country where a capital city swells up like an infected gland, attracting to itself manpower, which cannot be employed. Nor does it want to become a Marxist state.

Clearly Nyerere's philosophy of *Ujamaa* was more the quest for independence from colonial rule and influence than a belief in the imminent ecological disaster. Nevertheless, Nyerere's ideas of self-reliance resonated with the Blueprint's ideals for decentralisation, and also reaffirmed the possibility of a state-driven ecological agenda.

In pursuit of the ideals of the Blueprint, the world's first 'green' party, the Values Party, was founded in New Zealand in 1972. Unlike most environmental movements that focused mainly on conservation issues in the 1960s and 1970s, the Values Party 'provided a broader ideological critique of the state's preoccupation with materialist concerns' (Downes 2000, 474). The 88-page policy programme (Beyond Tomorrow), which was publicised as an election manifesto for the 1975 national election, committed the Values Party to limits on economic and industrial development, a stable population, liberal social policies, the devolution of power in the workplace and community, and environmental protection (Miller 1991). Essentially, the Party was responding to the exclusion of environmental interests from the political agendas that were dominated by the National and Labour parties. In particular, the Muldoon National Government's interests on big energy projects acted as a catalyst for the environmental groups' engagement with the state (Downes 2000). The successor to the Values Party, the Green Party of Aotearoa–New Zealand, which was formed shortly before the 1990 General Election, also adopted the same anti-industrialist stance.

Europe's first green party, the Ecology Party, held the view that the imminent ecological disintegration could only be averted through political transformation. In its view, that goal could be achieved by either acquiring power or by provoking other parties to take up the policies of the Ecology Party. Rudig and Lowe (1986) observed that the Ecology Party was formed by a core group of four people in 1973, as a result of the combined effects of the widely distributed doomsday predictions (such as the work of the Club of Rome) and the consequences of industrial decline in Coventry, England. The Party was therefore conceived in the fervour of the Blueprint, and presented itself to the public as 'part of a world-wide awakening to the need to reconstruct our way of life if there is to be any chance of survival with decency' (*New Ecologist* 1979, 21). It was guided by a belief that 'ecology lays special emphasis on the qualities of diversity, acknowledging and putting to good use the differences that exist between people and encouraging the widest possible variety of life styles rather than forcing everyone to accept the same mould' (Porritt 1979, 21). Whereas the Club of Rome was unclear about the sort of institutions that would be needed to achieve the equilibrium state, the Ecology Party reaffirmed the Blueprint's proposal for decentralisation. The Party's Manifesto for the 1979 General Elections called for setting up of Scottish and Welsh Assemblies and regional assemblies (*New Ecologist* 1979).[6] In this sense, the call for self-determination by the Ecology Party went beyond concerns with the environmental issues to suggest the political division of the United Kingdom into regions. Writing for the *Ecologist*, Waites (1984, 164) emphasised the need for regions in the United Kingdom in these words:

> Equally important, … and perhaps more elusive, is the recognition that our country must give more scope for regional consciousness not only in its everyday life but also in its political administration. We must encourage the regions more, not depress them further; we must find our strength from them, not look on them as a weakness; we must use them to renew our national consciousness for the soul of England is in its regions.

The Ecology Party emphasised economic self-reliance and political self-determination. Elsewhere other parties followed the same route of contesting elections.

According to Spretnak and Capra (1984), the Union of German Ecological Citizen's Group (*Bundesverband der Bürgerinitiativen Umweltschutz*) was the first German organisation to structure a grassroots movement, which promoted ecological consciousness on a large scale in the 1970s. At the time, ecological concerns in West Germany were shaped by internal and external influences. As we have noted above, the work of the Club of Rome, *The Limits to Growth*, had influence in different parts of the world, and West Germany was no exception. Inside West Germany, the post-World War II political and economic environment led to the emergence of a social movement, which sought to oppose the development of nuclear power plants, the use of missiles as part of the NATO defence system, toxic waste and the like; while, at the same time, promoting an alternative culture. With the influence of the counterculture in the United States and Britain, 'the alternative movement [in Germany] fostered multiple developments with similar intentions, rejecting forced adherence to any one dogmatic line of thought, a mode [it] associated with both the Nazis and the Marxist/student protests' (Spretnak and Capra 1984, 14).

The impetus for the formation of the Green Party in that country came from a group of ecologists, the Action Committee of Independent Germans (*Aktionsgemeinschaft unabhängiger Deutscher*, AUD), which urged the Christian Democrat parliamentarian and author of *A Planet is Plundered*, Herbert Gruhl, to join the AUD in order to launch a new party. Subsequently, Gruhl left the CDU to form his own group, Green Action Future. Groups such as the Action Third Way and the Free International University were also interested in the formation of a green party. The founding convention of the Green Party was held in Frankfurt in March 1979 and resulted in the establishment of the Further Political Association (FPA)–The Greens (Spretnak and Capra 1984). Two major issues at the convention were a nuclear-free Europe and a decentralised Europe of regions. The Green association entered electoral politics by contesting election to the European Parliament in June 1979. It surpassed the 5% threshold in the 1983 national elections and secured 27 seats in the *Bundestag* (Dryzek *et al.* 2003). Against this backdrop, the German Green Party combined civil protests and legal–administrative strategies to push for environmental changes through the political system. According to Dryzek *et al.* (2003, 38), 'the initial push for representation in the *Bundestag* was underlined by a belief that not only would this provide greater political power, but also that formal political channels would provide the movement with a complementary strategy to campaigning in civil society'. Of significance to the theme of this chapter is the founding of the Green Party under the combined influences of ecology and the post-World War II socio-economic developments.

Apart from participating in national elections, the Green parties have seized the opportunities of regionalism to develop an international profile and to participate in regional groupings such as the European Union in the form of a regional green movement. This is particularly true of the European Federation of Green Parties,[7] whose founding conference was held at Masala, Finland on 20 June 1993. The Federation reiterates the ideas of the MS in these words:

> The so-called progress of the past centuries has brought us into a situation where the basis of life on Earth is seriously under threat. While technological development may delay the deterioration of the environment for a time, it cannot prevent the ecological and social collapse of civilisation without a fundamental change in the ideology of unquestioned material growth, which still prevails.[8]

Unsurprisingly, the Guiding Principles of the Federation are based on three pillars, namely, ecodevelopment, common security and new citizenship. The Federation's notion of ecodevelopment is founded on sustainable development, opposition to a free market economy or state-controlled economy, and self-reliance within the context of regions[9] rather than national or administrative borders. It proposes an ecological world economy based on planetary solidarity and the protection of the diversity of ecological resources and the global commons as a prerequisite for the management of the global economy, and suggests the possibilities of imposing quotas on the extraction and use of natural resources. Two of the main security objectives[10] of the European Greens are fair distribution and sustainable use of natural resources, and the protection of the environment and biosphere. It subscribes to the post-Cold War concept of security, which considers the importance of social, economic, ecological, psychological and cultural aspects. The Federation calls for the New Citizenship, which could be realised through the decentralisation of power from the nation state to communities, districts and regions.

Though Green parties, as individuals or as a collective, have become a permanent feature of the political landscape (particularly in Europe), their record of performance in elections has generally been poor (Table 2.1), despite Müller-Rommel's contention that Green parties passed the threshold representation in 12 European countries between 1981 and 1997. Analysts ascribe the generally poor performance of Green parties to the ability of mainstream political parties to take on environmental issues and to make environmentally relevant political appointments. For example, the former US President Bill Clinton included environmental issues in his campaign, and appointed environmentalist leaders such as Gus Speth to his first administration (Dryzek *et al.* 2003). In places such as Norway, where environmental groups operate in close co-operation with the government, and where environmental considerations have largely been co-opted into the existing political system,[11] the Green Party has little chance of using environmental issues for electioneering. That is to say that the inclusion of environmental issues into the political agendas of the mainstream political parties denies green parties a niche in political elections.

Table 2.1 The performance of selected Green parties in the first five national elections.

Country	Total number of parliamentary seats	Number of Green Party seats in national parliament				
Austria	183	0	8	10	13	9
Belgium	212	0	4	9	9	17
Germany	520	0	28	44	8	49
Italy	630	13	16	11	21	17
Luxembourg	64	0	2	4	5	5
Sweden	349	0	0	20	0	18
Switzerland	200	6	11	15	8	9

Source: Adapted from Müller-Rommel (2002)

Clearly, the performance of Green parties cannot be measured effectively on the basis of environmental issues only. The lifespan[12] and organisation structures of Green parties and the political environment in which they contest elections all play a significant role in the performance of those parties. For example, the Green Party in Germany emerged from the confrontation between the environmental movement and the state, and its ability to develop a broad programme of societal change strengthened the party's position in electoral politics. According to Dryzek *et al.* (2003, 37), 'the political exclusion of the anti-nuclear movement [in Germany] helped fashion a strong oppositional counter-culture which went beyond specific policy goals to include issues of identity and alternative forms of actions and behaviour'. Nevertheless, the same cannot be said of the Green Party in Finland, which has the same duration in parliamentary representation as that in Germany (Müller-Rommel 2002). That is to say that the Green Party in Finland is a weak performer (Table 2.2).

Table 2.2 Electoral performance of the Green Party in Finland, 1987–2003.

Year	Number of elected representatives in Finland's political parties										
	SD	LA	NCP	CPF	SPP	L	TF	CDF	**GL**	PFP	EP
1987	56	20	53	40	12	0	9	5	**4**	n/a	n/a
1991	48	19	40	55	11	1	7	8	**10**	n/a	n/a
1995	63	22	39	44	11	0	1	7	**9**	2	1
1999	51	20	46	48	11	0	1	10	**11**	0	0
2003	53	19	40	55	8	0	3	7	**14**	n/a	0

Source: Finland (2004)

In the United States, the Greens' exclusion from the state has been self-imposed, not least because the question of whether Greens should be organised along party political lines has been a source of dispute among the Greens (Dryzek *et al.* 2003). The US Greens' first national impression in 2000 saw the Presidential candidate Ralph Nader and Winona Laduke gaining 2.6% of the national vote, and loosing the presidency to George Bush. The radical Green movement in that country has remained outside the arena of state politics and mainstream environmental movement, preferring instead to champion the environmental course at the grassroots level. The environmental justice movement has particularly been concerned with inequitable distribution of harmful wastes, particularly the concentration of those wastes in areas occupied by people of colour (Bullard 1990; Pulido 1996). While the environmental justice movement has successfully influenced the US environmental policy,[13] it has been less concerned with the intersection of ecology and human survival. That concern has been central to the other radical group, the bioregionalists.

Bioregionalism: the search for grassroots autonomous regions

In contrast to Green parties that sought to promote ideas of ecological integrity through electoral politics in Europe, the accentuation of ecology as the basis for societal organisation received a different emphasis in the United States. It took the form of a social movement that sought to protest and defend localities against the exploitation of human and natural resources and the consequent impoverishment (Aberley 1999). The different ideas and practices of the movement have collectively been considered as bioregionalism. Accordingly, Aberley (1999, 13) has described bioregionalism as 'a body of thought and related practice that has evolved in response to the challenge of connecting socially-just human cultures in a sustainable manner to the region-scale ecosystems in which they are irrevocably embedded'. Notably, bioregionalism, as a philosophy or paradigm, is largely an aggregate of different views and practices and is, as Aberley (1999, 13) put it, 'a story from many voices'. The doyen of bioregionalism, Sale (1985, ix), went to the extent of describing bioregionalism as 'fluid and organic, with perhaps as many approaches as there are bioregionalists', making it difficult, if not impossible for one person to speak on behalf of all bioregionalists. Nevertheless, both Aberley and Sale (1985) agree that there are common threads that cut across bioregionalists' views. According to Sale, the bioregional paradigm revolves around scale, economy, polity and society. These defining features of bioregionalism should be understood in the context of responses to ecological disintegration and the threats that that disintegration poses to the survival of species, including humans. As Sale (1985, x) wrote, bioregionalism is 'not merely a new way of envisioning and acting a very old American ideal, but is also a crucial, and perhaps virtually the only possible, means of arresting the impending ecological apocalypse'. Put in another way, the ecological apocalypse can best be arrested by using the appropriate scale at which environmental problems can be addressed and by restructuring society and the economy to reflect that scale. A discussion of each of these important elements of bioregionalism sheds light on the aspirations and goals of the movement.

Scales for ecological integrity

A strong body of work suggests that scale is important in addressing environmental problems, mainly because environmental problems cut across some of the most familiar scales, especially the nation-state. Consequently, solutions to those problems should occur at various scales, ranging from the local, through the national to global scales. However, bioregionalists emphasise one type of scale above others, preferring the use of the local scale at which people will probably feel connected to the problems at hand. In arguing for the bioregional scale – and criticising the call for environmental ethics – Sale (1985, 53) suggested that:

> people will do the environmentally 'correct' thing not because it is thought to be the moral, but rather the practical thing to do. That cannot be done on a global scale, nor a continental, nor even a national one, because the human animal, being small and limited has only a small view of the world and a limited comprehension of how to act within it.

Sale's suggestion is that humans will tend to protect the environment on which they depend. People who destroy their environment do so, because they are not living in the bioregional scale, but are instead serving distant economic forces. 'Dwellers in the land'[14] have a better understanding of how to live in that land and how to look after it. This view of scale and associated activities is common among bioregionalists, and is underscored by the belief that the earth is organised into natural regions – that vary in size – rather than artificial states.[15] In this sense, the natural region is the bioregion, which Sale (1985, 55) defined as 'any part of the earth's surface whose rough boundaries are determined by natural characteristics rather than human dictates, distinguishable from other areas by particular attributes of flora, fauna, water, climate, soils, and landforms, and by the human settlements and cultures those attributes have given rise to'. A bioregion, in this context, is 'a place and the ideas that have developed about how to live in that place' (Aberley 1999, 23), whose boundaries are first and foremost determined by the people who live within it, considering climatology, physiography, and animal and plant geography. That is to say that the boundaries of bioregions are defined by cultural and physical attributes, hence Brunckhorst (2002) considers bioregions as biocultural landscapes. Indeed, Gary Snyder, one of the earliest bioregionalists, propagated cultural and individual pluralism and suggested the division of human populations by natural and cultural boundaries (Aberley 1999). It should be emphasised that, for bioregionalists, the boundaries of a bioregion are best determined by the 'dwellers in the land'. Using examples in the United States, Sale (1985) argued that the Native American Indian habitation exhibited bioregional patterns, making the Native Americans true dwellers in the land, as well as being early and unconscious bioregionalists. They represent the single basic building block of the ecological world, the community. Sale's (1985) view is that bioregions are organised into a complex arrangement and, like Chinese boxes, are found within one another. In that arrangement, the ecoregion is the widest natural region within which other smaller bioregions such as georegion[16] and morphoregion[17] exist. In a different but related context, Lamela *et al.* (2005) have suggested the development of the science of world planning (geoism) that could be used for a worldwide territorial planning. Geoism seeks to combine knowledge on a global scale in order to co-ordinate environmental, economic, social and cultural policies for the attainment of sustainable development and to protect the wilderness, in particular (Lamela *et al.* 2005).

Though sharing the same concern over ecological disintegration and interests in the fusion between culture and ecology and institutional transformation, large-scale ecologists'[18] view of a bioregion is different from that of the bioregionalism movement referred to above. Like bioregionalists, ecologists acknowledge that 'humanity needs to learn to live within ecological laws that govern the capacity of the biosphere', but are opposed to the notion of a bioregion, which is 'the small-scale self-sufficient communities advocated by some "deep ecologists"' (Brunckhorst 2002, 4, 7). In representing the views of large-scale ecologists, Brunckhorst (2002, 8) referred to a bioregion as 'a regional-landscape scale of matching social and ecological functions as a unit of governance for future sustainability that can be flexible and congruent still with various forms of government found around the world'.

The differences between the bioregional movement and large-scale ecologists are found in the diverse roles that the concept of ecology play as an incarnation of either nature, science, idea or political movement, or a combination of some, or all, of these (see Haila and Levins, cited in Pedynowski 2003). Arguably, bioregionalism invokes ecology for clear political goals whereas large-scale ecologists accentuate the scientific understanding of ecosystems. In contrast to what amounts to the bioregionalists' local scale, ecologists' view of a bioregion is a regional landscape that mediates between local and global scales. The ecologists' logic is that the global and local scales are extreme and therefore less useful for dealing with critical functional problems, hence the need for a regional scale. This is underpinned by the idea that ecosystems operate properly, and that they can best be managed, at multiple scales. In support of the ecological notion of a bioregion, Brunckhorst (2002) has argued that a broader scale provides multiple options for resolving problems, and allows decision-makers to evaluate various components rather than deal with a single site. Practically, this translates into re-establishing large-scale ecological systems that amalgamate bioregions into ecoregions. Where natural ecosystems have been modified, ecoregions are established on the basis of the nature and causes of the spatial patterns that would have existed in the absence of disturbances. The guiding question for the establishment of ecoregions is: what would the ecosystem look like had there been no disturbance (i.e. in the ecosystem)? Bailey (2005) concedes that the primary problem of identifying ecoregion boundaries is to synthesise a variety of clues in order to arrive at reasonable boundary placement. In practice, different factors are used to establish ecoregions, depending on the nature of particular environments. Bailey (2005) has suggested that, where vegetation is used in defining ecoregions, late successional vegetation is more useful than pioneer vegetation; and potential vegetation is more useful than real or actual, as existing vegetation does not correspond well with the patterns of ecological regions. In spite of variations in the methods used, ecoregions cover extensive geographical areas, principally because the sub-division of the earth's surface into ecoregions is based on macroscale patterns of ecosystems (Bailey 2005).

Transformation of the polity

Bioregionalism dismisses the modern nation-state on the grounds that: (i) it imposes monoculture on people of diverse backgrounds; (ii) it has failed to address seriously the crisis of ecological peril and the political systems that have caused ecological disintegration; and (iii) that it is at odds with natural laws, which govern biotic communities without coercion, organised force or recognised authority. It proposes forms of human governance that are in line with those found among biotic communities. In support of this proposition, Sale (1985, 90) invoked the Chinese philosopher, Lao Tzu, to suggest that the best government is no government at all, arguing that 'a political vision based on the evident workings of the biotic world, would not celebrate centralised co-ordination, hierarchical efficiency, and monolithic strength … but rather, in starkest contraposition, decentralisation, interdependence, and diversity'. A bioregional polity would therefore replace the modern state through the promotion of diverse and decentralised institutions, which reflect the different cultures and how groups of people want to govern themselves. As Sale (1985, 108) put it, 'bioregional diversity … does not mean that every community … would construct itself along the same lines, evolve the same political forms … Truly autonomous bioregions would inevitably go in separate ways, creating their own political systems according to their own environmental settings and their own ecological needs'. In essence, this is a call for a variety of political systems that are not necessarily compatible or viewed as 'good'. Unlike the modern state that is based on common electoral norms governing the ruler and the ruled, decision-making in the bioregional polity is left to members of the community who are entrusted to do so in accordance with the existing cultural norms of that community. The size of the fundamental polity is estimated at 5000 to 10,000 inhabitants. It should be pointed out that the bioregional vision does not call for the establishment of isolated small polities, but cherishes the idea that decision-making and control of a bioregional polity should be at the local level, and that the polity may find it necessary to co-operate with neighbouring polities for the common good. Such co-operation may result in the formation of confederate bodies.

Bioregionalism not only emphasises the human interconnectedness with the landscape, but also informs transformative social change at two levels: 'as a conservation and sustainable strategy and as a political movement, which calls for the devolution of power to ecologically and culturally defined bioregions' (McGinnis 1999, 4). Rather than contesting national elections for purposes of decentralisation, bioregionalism propagates the creation of bioregions that would reintegrate human cultures into their supporting ecosystems. In essence, bioregionalism is a counterculture to industrial societies, proposing a form of decentralisation that can only be achieved by abolishing the modern state. The ultimate goal of bioregionalists is to establish some form of ecologically decentralised governance through both resistance to the destruction of wild systems and the homogeneity of national culture, on the one hand, and the renewal of the knowledge of how natural systems work, on the other hand.

Self-reliance: a bioregional economy

Bioregionalism proposes the phasing out of the capitalist economy, mainly because that economy emphasises competition, exploitation and individual profits; attributes that lead to the establishment of material prosperity for the few at the expense of the majority and the environment. An alternative and desirable economy is the one that conserves resources and the relationships and systems of the natural world, aiming at sustainability rather than growth. The bioregional economy is based on the principle of self-sufficiency, as in preliterate society. The bioregionalist view is that a self-sufficient bioregion would be economically stable, and free from external control and dependency and the structural flaws of foreign currency. Carr (2004, 70) has noted that 'the long-term goal of the bioregional movement is to establish bioregionally self-reliant economies of "place" networked geographically across "space" via communications and limited trade links.' In stark contrast to the neo-liberal agenda, which favours a single global market, the bioregional economy emphasises the view that the economy should be based on communal rather than private ownership of resources – the wealth of nature should be the wealth of all. Clearly, this is a radical view of the economy that chimes with a well-established critique of the capitalist economy.

The bioregionalism vision has been carried forward in various ways, including the creation of ecovillages, which Svensson (2002, 10) defines as 'communities of people who strive to lead a sustainable lifestyle in harmony with each other, other living beings and the earth'. They not only embody a way of living but are grounded 'in the deep understanding that all things and all creatures are interconnected', and 'represents a widely applicable model for the planning and reorganisation of human settlements in the 21st century' (Svensson 2002, 10). Unsurprisingly, the three main dimensions of ecovillages are ecological, social and cultural/spiritual. The ecological dimension is based on the concept of permaculture[19] and includes growing organic food, using local materials and traditions in building homes, adopting ecological business principles[20] and protecting and safeguarding biodiversity and wilderness areas. The social dimension of the ecovillage emphasises the values of living as a small community as opposed to the megastructures of the industrial world. The cultural/spiritual aspect of ecovillages fosters spiritual values and rituals that recognise and cherish the connection of human life and the larger cosmos. Examples of ecovillages that are based on principles of bioregionalism include the Sasardi Natural Reserve and Ecovillage (Colombia), Crystal Waters Permaculture (Australia), Svanholm (Denmark) and Lebensgarten (Germany) (Lindegger 2002; Madaune 2002). Clearly, ecovillages and the broad aims and practices of bioregionalism as a social movement provide the platform on which different interests can be pursued.

The coalescence of ideas and interests

The environmental movement has gone beyond the narrow concern with ecological integrity to include socio-economic issues that are of relevance and significance to the wide audience. These include the restructuring of the world economy, changing patterns of consumption, promoting equity and social justice, protecting the rights of indigenous people and women, changing the structural imbalances of trade, securing peace and nuclear disarmament, promoting democracy, opposing social exclusion and economic exploitation, and so on. In so doing, environmentalism has shifted the political realignment from the class-based left–right axis towards a materialist–postmaterialist axis (Miller 1991). Spretnak and Capra (1984, 3) observed that, 'in calling for an ecological, non-violent, non-exploitative society, the Greens transcend the linear span of left-to-right'.

Arguably, the bioregionalist critique of the modern state and the capitalist economy in the contexts of ecological disintegration does not only articulate the tenets of the Blueprint for Survival referred to above, but also provides a platform on which an alternative future is suggested and grounded on empirical and historical evidence. Bioregionalism is the latest reincarnation of centuries-old effort to define how socially just and ecologically sustainable human cultures could be created and sustained. It invokes the past in order to justify the need to redefine the future interrelationships between humans and their environment, advocating the view that human societies should live as 'future primitives' or that humans should be 'reinvented at the species level'. As Aberley (1999, 20) has observed, 'a deeply rooted respect for indigenous thinking and peoples is a tenet fundamental to bioregionalism', suggesting that 'the way to the future can be found by adapting genetically familiar ways of life practised by ancestors and surviving indigenous people' (Aberley 1999, 16). This is evident in Sale's (1985) assertion that one of the two great teachers of humanity is the traditional wisdom of mankind.[21] In fact, McGinnis (1999) has suggested that the idea of bioregionalism can be traced back to the native inhabitants of the landscape. In this way, bioregionalism reaffirms the history and culture of native people, echoing an anti-colonial stance, which has global relevance to previously colonised and marginalised societies. In the same logic, the European Federation of Green Parties calls for the recognition of human rights and the protection of the rights of minorities.[22] Moreover, the bioregionalist view that land should be owned communally is in concert with orthodox communism and is supportive to contemporary tenure systems in most parts of the non-western world, particularly Africa. Communal land ownership has generally been marginalised and was systematically denied a formal status throughout colonial rule. The point here is that bioregionalism's suggestion that indigenous knowledge and way of life is the perfect model for human survival chimes with the indigenous peoples' struggles for self-determination and their quest for resource rights, particularly land. In this way, bioregionalism contributes to the validation of indigenous knowledge systems.

The call for the transformation of the state runs through variants of environmentalism, gluing together different groups and institutions that are opposed to the workings and structures of the modern state. For example, ecologists are worried that most current institutions were developed to serve political goals that have very little, or nothing, to do with the sustainable management of natural resources and people at multiple spatial and temporal scales, and that electoral cycles produce governments with short-term planning (Brunckhorst 2002). Moreover, the structure of the modern state favours sectoral approaches that severely limit the government's ability to deal with environmental problems, which affect, or arise from, various scales. Over the years, the state has been seen as an enemy of the environment, mainly because of the negative impact of state policies on the environment and the apparent lack of political will to legislate a radical environmental course. The diminishing role of the state in the global capitalist world, its lack of capacity and financial resources, and the willingness of the private sector and individuals to carry the financial cost for conservation on behalf of the state and to profit from nature conservation initiatives, have contributed to the fusion of interests between bioregionalists, ecologists, capitalists, politicians and other social movements.

By reaffirming nativism and vehemently opposing the ecological destruction through industrialisation, bioregionalism appeals to a wider audience that does not necessarily subscribe to all bioregional principles and suggestions. Bioregionalists' opposition to the homogenising tendency of the high mass-consumption of the industrial society is in line with most tenets of contemporary anti-globalisation theorists and movements. Notwithstanding its supporters, who view it as a panacea for world poverty, globalisation has severely been criticised for increasing the poverty gap within and among nations and for creating a borderless world to the detriment of local cultures and institutions, including the nation-state. Bioregionalism's proposition for the re-establishment of ecoregions appeals to the scientific community, especially nature conservationists, who have been concerned with the appropriate scale at which biodiversity can best be protected and managed. The objective of the European Federation of Green Parties to bring at least 20% of Europe under some sort of conservation area and special programme area supports the ambitions of organisations such as the IUCN. That is to say that though operating from different standpoints, Green parties, bioregionalists and environmental NGOs share common features that not only define the relationship among them, but that also bring the wider environmental movement closer to most interest groups. Aberley (1999, 35) is of the view that the appropriation of bioregional values by mainstream institutions could be a threat to bioregionalism as the general public could identify with the rhetorical and pragmatic government-sponsored initiatives (devoid of bioregional values) 'rather than associate with bioregionalism with its grass-roots and organic origin'.

Environmentalism, by and large, promotes the creation of ecoregions. These are relevant to nature conservationists, not least because it is held that the term 'ecoregion' was first proposed in 1962 by the Canadian forest researcher, Orie Loucks (Bailey 2005). In fact nature conservationists have been concerned that national parks that protect remnants of biodiversity are too small and are biased towards certain species and habitats, thereby excluding critical elements of biodiversity. The limitations of national parks with regard to nature conservation are generally ascribed to the various purposes for which national parks were established to serve. These range from the use of national parks to establish strategic zones, build national identity, foster the separation of races, define state boundaries, and so on.

Ideas about bioregions and ecoregions are instrumental in the development of propositions that seek to preserve entire ecosystems and to establish corridors that link isolated pockets of existing and future protected areas (Vreugdenhill *et al.* 2003), and to bring these under a unified governance structure. Indeed, the combination of ecology and conservation accounts for the amalgamation of contiguous protected areas. In this sense, protected areas, particularly national parks, have been used in defining the boundaries of ecoregions, leading to the rise in TFCAs world-wide. The rationales for establishing TFCAs go beyond ecology to include the demand for institutional transformation. This demand is eloquently articulated in Brunckhorst's (2002) assertion that current institutions were not meant to manage natural resources and people in a sustainable manner, but were, instead, developed to serve different purposes. As governments develop short-term policies that represent election cycles, 'political power structure is not in place long enough to adapt' (Brunckhorst 2002, 48). Moreover, proponents of TFCAs view the compartmentalisation of the state as an impediment to the long-term management of ecosystems. The transboundary nature of environmental problems such as pollution, global warming, desertification and the like, together with the weakening of the state, have added weight to the call for the institutional transformation of the state. The solution to all these problems is seen as the establishment of a long-term power structure that not only transcends the election cycles of governments, but which also operates above the government. This view is in direct opposition to Green parties that want to be in charge of governments and the bioregionalists' quest for self-contained communities, but is supportive to positions that some environmental organisations have held since the end of World War II. The point here is that the role of non-state actors in propagating TFCAs has a long lineage, which can be traced back to the Club of Rome. In that lineage, the weaknesses of the state have repeatedly been exposed in order to either insert the alternative authority of non-state actors in transboundary natural resources or to restructure the state.

With regard to the theme of this volume, variants of the environmental movement are relevant to the creation of TFCAs, not least because they accentuate the sanctity of ecological systems while, at the same time, opposing the use of state boundaries as a tool for organising all human life and activities. This is succinct in Sale's (1985) claim that the first law of Gaia is that the face of the earth is organised not into artificial states but natural regions, which vary in size. As we shall see in Chapters 4 and 5, TFCAs are promoted on the basis of re-establishing ecoregions across state borders. Without pre-empting the discussions in subsequent chapters, TFCAs have conveniently adapted some of the guiding tenets of the environmental movement to conditions in Africa. The most obvious of these is the propagation of an environmentally friendly economy in the form of ecotourism, and the adoption of an ecoregion as an organising concept of space. In the absence of Green parties in Africa's political landscape, environmental NGOs have been at the forefront of TFCAs on that continent. In the chapter that follows, I discuss peace parks as a model for TFCAs in general, and in Africa in particular.

Endnotes

1 There are differences of opinion on the notion and causes of environmental crisis, and the meanings of that crisis have changed over time (see DeLuca 2001; Buell 2003; Hollander 2003).

2 The meeting was called by an Italian industrial manager and affiliate of Fiat and Olivetti, Dr Aurelio Peccei.

3 The first phase of the study was supported by Volkswagen Foundation.

4 It involves about 82 states, 111 government agencies, more than 800 non-governmental organisations (NGOs), and some 10,000 scientists and experts from 181 countries (www.iucn.org).

5 Including funding through bilateral and multilateral agreements.

6 Clearly, regionalism in the United Kingdom had political connotations, and has been central to the politics of the British Isles (see Jones and MacLeod 2004).

7 According to the website of the European Federation of Green Parties, the Federation had 31 member parties and seven applicant parties in 2005. The 31 member parties are Die Grünen (Austria), Bulgarian Green Party, Cyprus Green Party, Strana Zelenych (Czech Republic), De Grønne (Denmark), The Green Party (England and Wales), Eesti Rohelised (Estonia), Vihreä Litto (Finland), Agalev (Flanders), Les Verts (France), Green Party of Georgia (Georgia), Bündnis 90/Die Grünen (Germany), Politiki Oikologia (Greece), Zöld Alternativa (Hungary), Comhaontas Glas (Ireland), Federazione de Verdi (Italy), Déi Gréng (Luxemburg), Alternattiva Demokratika (Malta), De Groenen (Netherlands), Groen Links (Netherlands), Miljøpartiet de Grønne (Norway), Os Verdes (Portugal), Green Party (Rumania), Green Party of St Petersburg (Russia), Scottish Green Party (Scotland), Strana Zelenych na Slovensku (Slovakia), Confederación de Los Verdes (Spain), Miljöpartiet de Gröna (Sweden), Grüne/Les Verts/I Verdi (Switzerland), Ukrainian Green Party (Ukraine) and Ecolo (Wallonia). The seven applicant parties were the Green Party of Albania, Los Verdes de Andalucia, Green Union Armenia, Green Movement of Azerbaijan, Els Verds (Catalonia), The Latvian Green Party and Lithuanian Green Party.

8 http://utopia.knoware.nl/users/oterhaar/greens/europe/princips.html (accessed 21/11/2005).

9 In line with the perspective of Europe of Regions.

10 The other objectives are: to nourish freedom and democracy; to narrow the gap between rich and poor; creation of a pan-European security system; and general and complete disarmament.

11 Ardal (cited in Dryzek *et al.* 2003) is of the view that the two main parties in Norway, the Liberal Party and the Socialist Left Party, are sympathetic to the environmental agenda.

12 Müller-Rommel (2002) has classified Green parties according to their lifespan into three types, namely: Greens with a long duration of pre-parliamentary and parliamentary participation before entering government; Greens with long experiences of parliamentary participation and a low duration in pre-parliamentary activities; and Greens who had long extra-parliamentary experiences before entering parliament and government.

13 It persuaded the Clinton government to sign Executive Order 12898, which made environmental justice for minority and low-income populations 'a legitimate policy concern at the highest levels of the federal government' (Bowen and Wells 2002, 689).

14 In contemporary discussions about the people–environment relationship, the notion of 'dwellers in the land' is approximate to that of local communities.

15 The idea that the state is artificial has been used by proponents of TFCAs to suggest that state boundaries should be removed, because they are artificial.

16 Smaller bioregions within the large ecoregion, which are identified by clear physiographic features such as river basins and mountain ranges (Sale 1985).

17 A series of smaller territories within the georegion identified by distinct life forms, such as towns and cities.

18 Those propagating large-scale ecological units – see earlier views on this in MacArther and Wilson (1967).

19 'A philosophical and practical approach to land use integrating micro climate, functional plants, animals, soil, water management and human needs into intricately connected, highly product systems' (Bang 2002, 18).

20 Such as complementary currencies, alternative banks, and the expanded local informal economy.

21 Bioregionalism promotes the view that indigenous communities conducted their lives within the parameters of bioregionalism. Subsequently, this was constrained by population growth, new technologies, arbitrary nation/state boundaries, global economic patterns, cultural dilution and declining resources.

22 http://utopia.knoware.nl/users/oterhaar/greens/europe/principles.htm (accessed 21/11/2005).

Chapter 3

Peace through ecology? A soft approach to hard realities

> Linking up adjoining conservation [centres] between two or more countries restores the natural parts of Africa and enables animals to move in a greater area than before. The concept of transfrontier parks sends a powerful symbol that countries are ready to live in peace and solidarity (Nelson Mandela, cited in Godwin 2001, 30).[1]

> In addition to conserving biodiversity and contributing to tourism development and poverty alleviation, TFCAs also have the potential to promote co-operation between governments and promote a culture of peace and stability in SADC (Hall-Martin and Modise 2002, 11).

As the quotations above suggest, the amalgamation of adjacent national parks into transfrontier parks is touted as a symbol of peace between and among states. For this reason, transfrontier parks are also referred to as peace parks. The designation of these parks and the role they can play in promoting peace is generally traced back to the establishment of the Waterton/Glacier International Peace Park between Canada and the United States in 1932. The Waterton/Glacier Peace Park brought nature conservation to the forefront of international relations, leading to what Krasner (cited in Tanner *et al.* forthcoming) calls an international regime by which issue-specific relations among states are governed by means of principles, norms, rules and decision-making procedures rather than formal agreements such as treaties. Unlike contemporary peace parks in Africa that emphasise formal agreements and treaties, the Waterton/Glacier Peace Park remained an informal co-operative venture for almost seven decades.[2] It was formalised in 1998 when a Memorandum of Understanding was signed between Parks Canada and the United States Park Service (Tanner *et al.* forthcoming). Moreover, the establishment of biospheres in the two parks in the 1970s and their designation as a World Heritage Site in 1995 (on the criteria of natural area nominations) bolstered nature conservation in the Waterton/Glacier Peace Park.

Of significance to our discussion in this chapter is the use of the Waterton/Glacier Peace Park as a model for peace parks elsewhere, especially in Africa. Tanner *et al.* (forthcoming) have remarked that, as the world's first peace park, Waterton/Glacier was the forerunner of what would become a significant movement in conservation worldwide. Indeed, almost all discussions about peace parks take the Waterton/Glacier Peace Park as a point of departure (De Villiers 1999; Sheppard 1999; Ali 2002). The Rotary Club International, which lobbied for the establishment of the Waterton/Glacier Peace Park, regarded that park as 'a way to cement harmonious relations between [nations], while providing a model for peace among nations around the world'.[3] It could be argued that the establishment of the Waterton/Glacier Peace Park was, at the time, more circumstantial than the beginning of a concerted effort towards the establishment of peace parks across the globe. This is evidenced by the failure of the United States to effect similar strategies along the US–Mexico border in the south,[4] and the lack of interest in the creation of peace parks globally during much of the 20th century. Arguably, the Waterton/Glacier Peace Park suggests two important aspects of the peace parks model: the focus on international borders and the linkages between ecology and peace at the supranational scale. This chapter concentrates on the elements of the peace park model and the assumptions underlying its propositions. Proceeding from a general overview of the notion of peace parks, the chapter emphasises the application of a peace park model to the African contexts. Conceptually, cross-border peace parks are founded on the notions of ecoregions and carry the same message of peace that has been central to the environmental movement (see Chapter 2).

The peace park model: propositions and assumptions

The existence of different types of peace parks worldwide renders any attempt to generalise about them over-ambitious. In fact, the reasons for and the bases on which peace parks are established are generally different. Nevertheless, two main types of peace parks stand out; the first being parks that are established within particular countries as commemorative sites or as a way of preserving particular local sites. The second involves the amalgamation of contiguous national parks across state borders, using the need to preserve the ecological integrity of areas as a guiding principle.

Peace parks that are established at the intra-state level are linked to various projects, ranging from the creation of peace gardens – as in the public square in Sheffield (UK), Nathan Phillips Square in Toronto (Canada), the Sadako Peace Garden (California) – to establishing peace cities as in the case of Wagga Wagga[5] in Australia. There is a general trend towards establishing intra-state peace parks as commemorative sites. Examples of these include the Hiroshima Peace Park (Japan) and the Gallipoli Peace Park (Turkey). Canada's ambitious peace park project involved dedicating a park for peace by each of the 400 cities and towns in that country as part of the commemoration of Canada's 125th birthday as a nation in 1992. In its pursuit of locally-based peace parks, the US-based International Institute for Peace through Tourism has planned to establish peace parks throughout the United States by 2010. On the eve of the 21st century, a Global Peace Parks Project was launched with the aim of establishing 2000 peace parks, with at least one park in every nation (International Institute for Peace Through Tourism 1999).

With the exception of initiatives such as the establishment of the Dag Hammarskjöld International Peace Park[6] in Zambia, most of the commemorative parks in Africa are hardly peace parks. Their designation generally reflects memories of struggles against colonial rule in the continent. For example, South Africa opened the Freedom Park[7] in the capital, Pretoria, in 2004 to mark the country's 10th anniversary of its liberation from apartheid. The central theme of the park is the struggle of humanity for freedom and dignity in South Africa. According to the Freedom Park Trust (n.d., 1–2), 'the Freedom Park must stand as a monument to [the] new democracy founded on the values of human dignity, equality, human rights and freedom. A symbol of the tortuous journey to, and sacrifices made for freedom. A celebration of the achievement of democracy and a beacon of hope and inspiration for the future'. Elsewhere, in Swaziland, King Sobuza II Memorial Park has been created as a tribute to the King's contribution to the independence of that country.

Though commemorative parks referred to above provide a useful background for reflecting on peace parks, they do not offer insight into the establishment of peace parks between and among states. That insight is gained from the second type of peace parks, those that are established across political borders. It is from these cross-border peace parks that this chapter seeks to understand the international peace park model. The choice for the cross-border peace park model has been made on four main grounds, the first being that international bodies have endorsed them as an ideal type of peace parks worldwide. Secondly, they are currently the main, if not the only, type of peace parks being promoted on the African continent. Thirdly, cross-border peace parks in Africa have gained currency among analysts, politicians, consultants, NGOs and local citizens. Fourthly, and more importantly, cross-border peace parks are central to the theme of this chapter and the volume as a whole.

It would be naïve to assume that peace parks have the same meanings worldwide, more so because, as I have intimated above, there are different ideas and examples of peace parks. Even peace parks that cross state borders have been subjected to different interpretations, in part because of the perspectives that have been adopted in analysing them (Chapter 1). The major weakness in most of those perspectives, though, is that the propositions on which the notion of peace parks is founded are often glossed over, and have therefore not been subjected to rigorous analysis. The identification of, and insights into, those propositions are imperative, not least because the propositions for, and assumptions[8] about, peace parks are critical to our understanding of peace parks. Our failure to understand the assumptions made about peace parks – by proponents of peace parks – is bound to weaken our assessment of these parks. At the level of abstraction, understanding the propositions and assumptions related to peace parks is helpful for analysing the ways in which peace parks are framed and how they operate on the ground. Against this background, the discussion below is an attempt to make a contribution towards a systematic exposition of propositions on peace parks and to suggest research themes covering the language of peace parks.

Proposition 1: International boundaries are more important for peace parks than sub-national ones

The use of concepts such as transboundary protected area (TBPA), transboundary conservation area (TBCA) and transboundary natural resource management area (TBNRM) has generated a debate about the types of boundaries referred to in these descriptions. Clarity on the types of boundaries is considered important for the designation and certification of peace parks. The certification process, in turn, requires the standardisation of boundary terminologies. Hanks (2000, 3) captured the need for such a standardisation in the following words: 'the development of internationally recognised categories, principles and criteria for peace parks is an essential prerequisite for the adoption of uniform standards and international institutional frameworks for joint management and integrated planning'. The organisation for protected areas in Europe, the EUROPARC Federation, has developed a certification process[9] for transboundary parks by which the quality of co-operation between parks is examined in five main areas: nature and landscape conservation; education and communication; recreation and sustainable tourism; research and monitoring; and mutual understanding and the promotion of peace (EUROPARC 2003).

The main concern among proponents of peace parks is whether the boundary should be scaled at the national or sub-national levels, given that states exist in different forms and sizes worldwide. In other words, the standardisation of boundary terminologies in peace parks is bound to confront challenges arising from different boundaries such as those of federal, unitary and marine states, as all types of states are called upon in the service of establishing peace parks. The axis of the debate about the scale at which boundaries should be recognised for purposes of peace parks has succinctly been captured in the criteria that have been suggested for designating peace parks, namely, that, 'there should be at least two protected areas, as defined by IUCN, sharing a common national or sub-national boundary' (Sandwith *et al.* 2001, 5). The implication here is that peace parks can be established between and within states. Sub-national boundaries have been included in the criteria to accommodate federal states, and are considered relevant to situations in which there is a need to lobby local institutions and other land users for purposes of establishing peace parks. This is particularly true where the expansion of protected areas can logically be achieved through encouraging landowners such as farmers to participate in nature conservation by joining their farmland to adjacent protected areas. As I will show in Chapter 6, there are concerted efforts towards incorporating communal lands and private farms into peace parks in most parts of southern Africa. The use of sub-national boundaries in the creation of peace parks is also supported on the basis that those boundaries are important for increasing the number of peace parks.

Despite those arguments, establishing peace parks across sub-national boundaries has received little support among advocates of peace parks, because of a number of reasons. The first one is that peace parks are aligned with the nature conservation paradigm that emphasises establishment of protected areas across state borders. Secondly, donors are increasingly becoming less interested in nature conservation projects at the intra-state level, and are therefore more favourable towards international peace parks. For example, the three main environmental NGOs, namely, Conservation International, WWF and the Nature Conservancy, are more interested in large-scale conservation strategies than in conservation projects at the sub-national level (Chapin 2004). Thirdly, the hype about existing and potential transboundary protected area complexes is based on international rather than sub-national boundaries. Zbicz and Green (1997) estimated that there were 136 transboundary complexes for protected areas, the majority (45) of which were in Europe, followed by Africa with 34. The fourth reason is that one of the main supporters of the peace parks initiatives, the tourism industry, is exclusively interested in peace parks across state boundaries. The International Institute for Peace Through Tourism (IIPT) that was founded in 1986[10] promotes the idea of peace through international tourism. In its Amman Declaration of 2000 it committed itself to building the culture of peace through tourism, espousing the view that travel and tourism is one of the world's largest and fastest growing industries, creating one in 11 jobs, contributing to international and regional economic growth, bridging disparities between developed and developing countries, and bringing prosperity which fosters peace. It should be emphasised that the notion of peace through tourism is biased towards international tourism; the implication being that the tourism industry gives strong support to peace parks across state borders (see Chapter 7). The fifth reason is the recent upsurge in ecological research that supports bioregional planning across state boundaries (Chapter 2).

It follows that the international boundary dominates conceptions of and practices around peace parks. It has been argued that to use TBPA/TBCA/TBNRM to refer to situations across any boundary is counterproductive and that the notion of the boundary should reflect the paradigm shift towards conserving biodiversity between two or more states.[11] Indeed, the following definition of a peace park favours the international boundary:

> A peace park is a formally gazetted transfrontier complex, involving two or more countries, which is under a unified system of management without compromising national sovereignty, and which has been established with the explicit purpose of conserving biological diversity, encouraging the free movement of animals and tourists across the *international boundaries* within the peace park, and the building of peace and understanding between the nations concerned. (Hanks 2000, 4)

Hanks (2000) further suggested that one of the minimum requirements for a formal listing of an area as a peace park should be the participation of at least two sovereign countries. It was anticipated that establishing the requirements for the formal recognition of peace parks would assist in developing a category for peace parks similar to IUCN categories for protected areas. Influential figures on peace parks in southern Africa such as John Hanks[12] and Leo Braack[13] have favoured the use of international rather than sub-national boundaries in the creation of peace parks, a factor which accounts for the model of peace parks that has been promoted in Africa. The adoption of an international context of a boundary was also defended on the basis that it would garner political support and international donor funding. The support for international boundaries also comes from guidelines and manuals that have been developed since the 1990s to assist in establishing TFCAs. More crucially, proponents of peace parks have capitalised on existing international conservation instruments such as Ramsar, CITES and World Heritage Convention to push for the use of international boundaries in TFCAs.

The choice of borderlands as sites for peace is underpinned by the classical view of borders as zones of friction and war, hence the need to change their functions. As I will show later in this chapter, the attempts to change the functions of borders on ecological grounds for purposes of peace is misplaced, because the origins and functions of borders are inextricably linked to national, regional and international complexes. Moreover, disputes over borders have their own peculiar histories that cannot be subjected to general and common solutions.

Proposition 2: Boundaries around peace parks are natural

The advent of international peace parks has been followed by the emergence of a hegemonic boundary discourse in which national boundaries are presented as an irrational and artificial human creation, which should give way to 'natural boundaries'. In that discourse, 'natural boundaries' are equated with the limits of ecosystems, and some species have been used to define those boundaries. For example, Sheppard (cited in Fall 2005, 114–115) asserted that 'important species, such as the Ibex in Europe and the Cougar in Central America are not interested in artificial boundaries drawn on maps.[14] Protected areas, reflecting this perspective, must broaden their outlook beyond traditional boundaries if they are to survive in the next century'. As we shall see below, the migratory routes of elephants have been used to define the boundaries of TFCAs in Africa.

Peace parks are founded on the view that ecosystems that had been divided by artificial – mainly state – boundaries should be managed as a single ecological unit defined by 'natural boundaries'. The cliché that 'nature knows no boundaries' has thus been used to frame the main function of political boundaries as that of separating that which should naturally belong together. Reference to the boundaries of peace parks as 'natural' serves to confer their status as part of creation. In this sense the Rotarian John Bohlinger remarked that 'God himself spent a little more time when creating the majesty of the Waterton–Glacier [International Peace Park],[15] as if the two national parks were predestined to become a peace park. State boundaries are not inherently disintegrative. Apart from designating an area over which the state exercises its sovereignty, the boundaries of the state were also meant to promote national unity and identity. Moreover, boundaries do not only separate population groups but also mediate contacts between them (Nugent and Asiwaju 1996), and their functions change over time (Paasi 1996; Anderson and O'Dowd 1999; Glassner and Fahrer 2004). It should be emphasised that political boundaries have also played an integrative role in other contexts, such as in the formation of regional blocs. These functions and qualities of boundaries have been marginalised in the boundary discourse of peace parks in order to promote the view that national boundaries are inherently divisive.

Equally problematic is the notion that peace parks follow nature's design (EUROPARC 2003), because there are different interpretations and representations of 'nature' (see Coates 1998; Castree 2001). Arguably, peace parks are founded on the separation of humans from their biophysical environment and perpetuate the idea of the wilderness (Ramutsindela 2004b). The assumption that the boundaries of peace parks are determined by nature (Turnbull 2001) disguises the fact that measuring nature's design, including the interpretation of that design, is another human activity that reflects the interests and ambitions of people and institutions involved. Paradoxically, the criticism of national boundaries as an outcome of a political process ignores the fact that the establishment of peace parks – at least in Africa – is also accomplished through yet another political process, which culminates in the signing of agreements and treaties. The very act of establishing peace parks is an expression of political power over certain spaces and species – it establishes a regime, which determines the governance of, and access to, natural resources. Accordingly, Fall (2005) has argued that the establishment of a transboundary entity on the basis of several existing protected areas implies a process of reterritorialisation by which a common reinvention and redefinition of both social and spatial practices take place. The creation of peace parks as a space was facilitated by the redefinition of environmental problems on both sides of the border as the same, irrespective of the different contexts from which those problems arose. In Africa this process was enhanced by a strongly held view of Africa as a homogenous place. I argue that the existence of protected areas adjacent to each other does not necessarily mean that they share the same problems,[16] as advocates of peace parks would like us to believe.[17] Protected areas do protect 'different things' and the countries in which they are located are often facing different socio-political and environmental challenges. Even the presence of the same animal in different countries could assume different environmental meanings. Fall (2005) has hinted that in the Maritime Alps, the return of the wolf was considered a problem in France but not in Italy.

It follows that understanding the notion of the boundary in peace parks demands the application of broad theories of boundaries that go beyond essentialist interpretations of boundaries. There should be a recognition that the construction of boundaries at all scales and dimensions takes place through: narrativity (Newman and Paasi 1998; Waters 2002); manifest power dynamics (Foster *et al.* 1988; Painter 1995); are part of material and discursive processes (Paasi forthcoming); and encompass not only the formal frontiers between states, but are also bound to the whole idea of territoriality (Clapman 1996). It could be argued that peace parks are part of the 'boundary-producing' process by which the state is brought in to assist in the redefinition of the sanctity of its borders. The question that arises from boundaries around peace parks is whether the impact of those boundaries on local residents would be any different from that of existing national parks.

Proposition 3: The environment is indissolubly interlinked with instability and peace

Ali (2002) believes that the decades-long war between India and Pakistan can be resolved through the establishment of the peace park over the Siachen region. Inspiration for establishing such parks has been drawn from examples such as the Peru–Ecuador Peace Accord[18] (1998) and the La Amistad protected area between Costa Rica and Panama (Sheppard 1999).[19] The notion of peace parks has therefore added a new dimension to the peace movement, which has a long pedigree within the broad environmental movement (Chapter 2). A brief discussion of the peace movement is useful for understanding peace parks in the contexts of security and peace.

Various strands of thought about peace are rooted in opposition to war as evidenced by pacifism, humanism and rationalism; and there are continuities between past and present peace actions. Carter (1992) observed that the first peace societies, such as the American Peace Society,[20] were founded on the belief that states could settle disputes on rational discussion without recourse to war, and that those societies drew upon a much earlier commitment to oppose war, mostly on religious grounds.[21] That is to say that peace was generally defined in terms of avoiding war between or among states, hence the Free Trade Movement of the 1940s 'believed that trade would create such a strong economic interest in peace that even governments adhering to obsolete attitudes would be prevented from launching wasteful and destructive wars' (Carter 1992, 4). Although the first organised peace groups were founded after the Napoleonic Wars and had their roots in Britain and the United States, World War I fundamentally shaped the direction of peace movements worldwide. This was so because of the consequences of the war and the increased public commitment to avoiding future wars through disarmament. For example, World War I saw the emergence of the Women's International League for Peace and Freedom, the *Esperanto* for Peace, the International Fellowship of Reconciliation and so on. The focus of peace movements on disarmament was seriously challenged by the outbreak of World War II and the subsequent military built-up of the Cold War. After World War II, scientists lobbied for control of atomic energy, resulting in the Baruch Plan of 1946, which proposed the formation of a supranational authority to monitor all nuclear plants (Carter 1992). Following the H bomb test by the United States in March 1954, and the health risks from radioactive fall-out, most peace movements campaigned against nuclear testing. Opposition to nuclear testing in the United States led to the formation of the National Committee for Sane Nuclear Policy in the late 1950s and the National Council for the Abolition of Nuclear Weapons Tests.

Of relevance to the discussion of this chapter is the link between peace movements and the broad environmental movement. According to Carter (1992, 108–109), 'the most direct link between the new Green movement and campaigners against nuclear weapons has been concern about the dangers of nuclear power'. That is to say that the growing environmental concerns of the late 1970s accounted for the strong public opposition to nuclear weapons. The Greens were worried about the mining and transport of uranium, and used their scientific expertise to oppose the use of nuclear power. At the time, circumstances were propitious for such campaigns, because the Cold War, particularly in the 1970s, raised the fear of a nuclear holocaust. Groups such as Greenpeace organised protest voyages to testing zones and found common cause with other peace movements. Notwithstanding responses to environmental damages associated with the Cold War, the intensity of that war in the 1980s opened up the possibilities for reformulating the term 'security'. Unsurprisingly, the Independent Commission of Disarmament and Security Issues mooted the idea of a 'common security', which placed environmental issues on the agenda of international politics in the late 1980s.

The end of the Cold War in 1988/1989 did not only result in fading public fears of a nuclear holocaust, but also robbed the superpowers of their justification for the control of client states. In the absence of superpower control, the concern was that conditions for instability that had been suppressed by the superpowers would lead to the eruption of conflict.[22] Moreover, the armies and institutions that had been put in place to wage the Cold War became redundant as a result of the end of that war. Consequently, environmental protection became a new assignment for the armies in developed countries (Brock 1991). Arguably, environmental issues were used to redefine the notion of national security to include areas beyond the borders of the state. Thus, the looming disaster of the nuclear war was replaced by the disaster that would be triggered by the depletion of the environment and new notions of security were required to reflect that change. In support for a redefinition of security in the context of environmental threats, Matthews (1989) argued that in the same way that nuclear fission was the dominant force in security strategies during the Cold War, the driving force of the coming decades would most probably be environmental change. Dalby (1992, 503) noted that 'the juxtaposition of security and environment [came] at a time when the conventional understandings of the former [were] no longer so obviously politically relevant while concern with the latter [was] much in public view'. In the United States the environment was considered an element of the National Security Strategy in 1991 (Butts 1994).[23]

Subsequently, efforts were made to define the US Department of Defense's environmental security role on the basis that the stress from environmental challenges contribute to political conflict, which requires the Department of Defense to mitigate the impacts of adverse environmental actions leading to international instability (Matthews 1989; Butts 1994). It was suggested that a US-led environmental security mission would give NATO a visible and relevant role after the waning of the Soviet Union threat. Environmental challenges that undermine the legitimacy of governments and create civil strife or harm US strategic alliances were considered a threat to US national security interests.[24] It was therefore recommended that environmental security should be established as a principal objective of the National Security Strategy and that it should be included in National Security Council and foreign policy planning (Butts 1994).

Notwithstanding all these developments, the link between environment and security is not readily clear, simply because environmental issues are neither given nor are they viewed from a common perspective. Scientific research, donor interests, NGOs, politics and so on, variously influence the ways in which environmental issues are conceptualised and dealt with. Moreover, unlike military threats that are focused and amenable to straightforward responses, environmental threats are not easily compartmented (Deudney, cited in Dalby 1992). The security–environment nexus has aptly been questioned in this way: 'given that security is often understood in terms of stability and the perpetuation of the current social and economic arrangements then the environmental agenda can, and often is, seen as a threat to national security' (Dalby 1992, 511–512). Despite these criticisms, advocates for peace parks insist that there are linkages between peace and the environment and that ecological co-operation could develop into an independent variable in international relations. In pursuit of this view Brock (1991) proposed four linkages between peace and the environment:

(a) *Causal relationship*
The negative relationship between peace and the environment is that the use and control of natural resources have contributed to conflict, violence and war. On the positive side, environmental pressure to reduce violence could generate strong opposition to war. Brock (1991, 411) hypothesised that, 'if the environmental repercussions of military activities cannot be confined to war area, international pressure may mount to avoid wars, for ecological reasons', as was the case with resistance against the war in Kuwait in the early 1990s.

(b) *The instrumental linkage*
The environment could be manipulated for hostile purposes, as was the case with the Vietnam War in which the United States used chemicals to denude trees in order to expose its target. Nevertheless, there is a view that the environment could be used to bridge conflict. According to Brock (1991), environmental co-operation for the protection of the Baltic Sea was instrumental in bridging the East/West conflict.

(c) *The definitional linkage*
Environmental degradation can be defined as war on the basis that the effects of environmental degradation are the same as those of military wars. Moreover, since humans are part of nature, peace with nature implies peace among people (Brock 1991).

(d) *The normative linkage*
There is a view that environmental degradation constitutes a non-military threat to the territorial integrity of the state and its economic well-being. Here environmental security is aligned to the idea that 'traditional security policy enhances overall insecurity and that this contradiction can be overcome only by pursuing a policy that defines the security needs of the one party in terms of the security needs of all parties' (Brock 1991, 417).[25]

The linkages between ecology and peace are therefore based on the assumption that ecological disintegration, if left unattended, will lead to fierce competition over scarce resources, thereby becoming a recipe for war.[26] Indeed, suggestions have been made that future wars will be fought over scarce water resources. In the same vein, the Nobel Peace Prize winner, Wangari Maathai (2005), has emphasised the point that fighting between warring farmers could be linked to environmental issues, especially the scarcity of resources. A counterargument to this view is that resource abundance can equally contribute to conflict. Drawing from literature on armed conflicts, Le Billion (2001, 563) has argued that, 'contrary to the widely held belief that abundant resources aid economic growth and are thus positive for political stability, most empirical evidence suggests that countries economically dependent on the export of primary commodities are at higher risk of political instability and armed conflict'. At the local scale, the creation of protected areas has often led to conflict between conservationists and local populations. The interpretations of the links between the environment/resources and conflict also account for different views on conflict mitigation and resolution strategies. Of relevance to our discussion in this chapter is the proposition for environmental peacemaking, which is premised on the view that certain attributes of environmental concerns could bring acrimonious parties to some sort of co-operation, and can be used as both entry and exit points between adversaries (Ali forthcoming). For this to happen, parties should be provided with a clear picture of a common environmental threat so as to highlight the importance of co-operation to alleviate that threat.

It follows that environmentalists have used both the effects of war on the environment and the framing of environmental threats in their peace campaigns. For example, in 2001 the IUCN (2001) proposed that TFCAs should be used for peace and co-operation. I argue that, by appealing to the morality of peace, proponents of peace parks render any opposition to the establishment of such parks as an anti-peace effort. The notion of peace parks is vaguely promoted as one of the distinctive aims of transfrontier conservation, yet no single treaty or Memorandum of Understanding (MoU) for TFCAs in southern Africa has the promotion of peace as one of its objectives.[27] Carter's (1992, 13) analysis of the distinguishing features of peace movements is instructive: 'peace, like democracy is an attractive label, which may be appropriated for propaganda purposes. The goal of peace is, moreover, too general to be itself a distinctive aim which separates peace groups from everyone else since most people desire peace for much of the time'. An understanding of the roles of peace parks in the promotion of peace requires that we should know what is implied by 'peace' in peace parks, the causes of conflict and violence and how peace parks respond to them, and the location of peace parks in relation to zones of political conflict. Advocates of peace parks are the first to admit that it is difficult to measure the impact of peace parks on the peace process, mainly because of the interplay of multiple factors in that process. It has been suggested that the Peace and Conflict Impact Assessment tool, which is applied to development and humanitarian interventions, is relevant to assessing and monitoring peace in peace parks (Besançon and Hammill forthcoming). My point here is that, though the argument for peace parks is that they provide an added value to the peace process, the impact of peace parks on that process is unclear. The added value from peace parks makes it almost impossible to assess the usefulness or otherwise of peace parks towards achieving the goal of peace on which they are founded.

In summary, peace movements are by and large preoccupied with states and their behaviour; promoting the view that peace can be achieved through engaging states at the supranational level. This view informs most peace-building strategies, including the establishment of peace parks. In no way do I suggest that all peace parks have been established across the borders of nation-states; there have been initiatives to create peace parks as sites of memory as we have noted in the beginning of this chapter. Rather, it is to acknowledge that the notion of peace parks is principally promoted on the basis of the need to resolve conflicts between and among states. In Africa, such a state-centric view has been used to frame problems facing African states in order to market the notion of peace parks.

Reinterpretations of Africa's problems for insertion of peace parks

Carter (1992) has mooted the idea that peace campaigns can be classified in terms of what they are against and that an important criterion for identifying peace organisations is that they are autonomous from government or political parties. Like the anti-war and anti-nuclear testing peace movements, proponents of peace parks have developed a target of what they are against, namely, environmental threat. As I have intimated in Chapter 2, environmentalists view threats to the environment not only in terms of environmental harms, but also frame those threats in the context of human survival. Arguably, environmental threats are key to the conceptualisation and marketing of peace parks in Africa. Those threats – and their relevance to peace parks in Africa – can more appropriately be understood in the contexts of the propositions we discussed above.

The proposition on the link between the environment and peace has its peculiar character in Africa for reasons that will become clear in a moment. The end of the Cold War and white minority rule in South Africa had security implications for southern Africa, a region that was to become a laboratory for peace parks. In practice, the end of the Cold War resulted in diminishing military support to southern African states by countries such as Russia and Cuba. Equally, the end of apartheid rule was followed by South Africa's abandonment of its regional destabilisation policy[28] and its nuclear weapons programme in the early 1990s.[29] The combined effects of these global and national developments were the proliferation of private security services in Africa (Mills and Streamlau 1999) and a wide-ranging demilitarisation process in southern Africa (Williams 1998). Notably, these developments posed new security challenges, leading to a profound shift in ideas about security in the region and the continent. New ideas about security in the region did not only emerge at the same time as those in the only remaining world superpower, the United States, but were also couched in post-Cold War terms. For example, the Deputy Director in the South African Defence Secretariat, Van Vuuren (1998), ascribed the increasing post-1989 turbulence and military operations to the loss of superpower influence over alliances and developing nations. The implication of this perspective is that the region required a new security policy in line with the national, regional and global security demands. Cawthra (1997) noted that the evolution of security policy in South Africa and the wider SADC region in the 1990s was underpinned by the need to develop a development-oriented approach in which it was assumed that contemporary threats are non-military and transnational. Whether the new policy was a result of a change in mindset among the ruling elite or an outcome of external pressure is debatable.[30] Nevertheless, there is evidence that profound changes occurred in security thinking at different levels in the 1990s.

It could be argued that, as elsewhere, environmental issues were incorporated into the reformulation of security thinking in the region, as evidenced by academic writings and comments from the military establishment. For example, the *Editorial* of the African Security Review (1998) considered environmental degradation as one of the salient threats to human life in Africa.[31] Following the same line of thinking, Steven Metz (2000) of the US Army Strategic Studies Institute, warned that many African societies face a bleak future not only from social unrest and violence but also from challenges posed by the natural environment. This is in line with post-Cold War security thinking, which holds that changes in the environment impinge upon security issues. By way of example, desertification leads to low food production and therefore to starvation and social unrest. Accordingly, Metz (2000) suggested that solving Africa's environmental problems will contribute to the security of the continent, and also promoted the view that Africa's strategic vision should include regional and subregional co-operation on shared problems. In practice this meant that the military should contribute to regional security by participating in cross-border development projects. To this end, the Director of Operational Policy in the South African Defence Secretariat, Williams (1998, 33), argued for the need to reduce excessive military expenditure and the reallocation of funds to development, which he defined as 'a marked and sustainable improvement in the economic outputs of a country, its employment levels, its political and social stability, its psychological well-being, and its level of cultural and spiritual enrichment and harmony'. The view that solving environmental problems is an important step towards peace is supportive of the notion of peace parks. However, proponents of peace parks have replaced the language of peace and security with that of peace and co-operation, thereby avoiding damaging political consequences that are often associated with the notion of security. Clearly, any reference to peace parks as an instrument of national security would potentially be detrimental to marketing the idea of peace parks to African governments. This is so, because the overt links between national security and peace parks would have been highly problematic. It would have given rise to the pertinent question of whose security is at stake. Unsurprisingly, the language of peace and co-operation is common among cross-border conservation projects of all sorts. In the construction of transboundary spaces co-operation is presented as inherently good and is theorised as the modification of the functions of boundaries (Fall 2005).

Co-operation for peace or peaceful co-operation?

As I have hinted above, advocates of peace parks argue that certain attributes of the environment, especially those related to environmental harms, provide the platform for dialogue between adversarial states. This logic is questionable in Africa where the environment has not played a significant role in the definition of peace and security as evidenced by the Protocol Relating to the Establishment of the Peace and Security Council of the African Union. The Protocol emphasises the concerns about the prevalence of armed conflicts in Africa and the resulting refugee problem, the scourge of landmines and the impact of the illicit small arms. The environment (as a word or concept) has not been mentioned in the 11 Principles guiding the Protocol. Instead, Principle (d) refers to the interdependence between socio-economic development and the security of peoples and states while Principle (i) upholds the respect of borders inherited on achievement of independence (African Union 2002). What becomes handy for peace parks' advocates is the description of the first objective of the Peace and Security Council as to 'promote peace, security and stability in Africa, in order to guarantee the protection and preservation of life and property, the well-being of the African people and their environment, as well as the creation of conditions conducive to sustainable development' (African Union 2002, 5).

In the context of peace parks in Africa, the concept of sustainable development is interpreted to mean the protection of biodiversity while allowing the controlled use of biodiversity for socio-economic development to ensure the security of, mainly, biodiversity. In this sense, security of biodiversity translates into war for biodiversity, which is justified on moral grounds in most parts of Africa (Neumann 2004). Notably, private military firms have been contracted by organisations such as WWF to conduct anti-poaching operations in Equatorial Guinea (Smith 2005),[32] while defence forces have been used to carry out similar tasks in Zimbabwe, Kenya and Tanzania, for instance. Though these measures were taken with regard to security of biodiversity in national parks, the use of national parks as the core areas of peace parks[33] implies that the notion of security in national and transfrontier parks remains basically the same.[34] In this sense, the focus of security is more on biodiversity than on humans. In fact discussions about security in peace parks are biased towards seeking mechanisms for dealing with crimes such as poaching, drug trafficking and illegal immigrants.

In practice, the idea of peace implied in peace parks has been reduced to collaborative management. Evidently, the objectives of peace parks in southern Africa have been described as 'to consult and co-operate with each other' (General Trans-Frontier Conservation and Resource Area Protocol), 'to foster trans-national collaboration and co-operation among the parties' (Treaty of the Great Limpopo), 'to promote the co-ordination of the management, control and development of the parks (Bilateral Agreement on the Kgalagadi Transfrontier Park), 'to foster trans-national collaboration and co-operation between the parties' (Treaty of the Ais-Ais/Richtersveld Transfrontier Park), 'to promote cross-border co-operation and interaction' (Lubombo Ndumu–Tembe–Futi Transfrontier Conservation and Resource Area Protocol) and 'to facilitate co-operation' (MoU on the Maloti–Drakensberg Transfrontier Conservation and Development Area).

To assume that co-operation through peace parks constitutes the basis for dialogue – and therefore peace – among African states is to ignore the fact that most conflicts and violence in Africa are intra-state rather than inter-state in nature (Cawthra 1997).[35] More importantly, the assumption underplays the impact that existing formal agreements have on the continent. Stated differently, how would peace parks promote dialogue among African states in ways that are dissimilar from other equally important existing agreements, and what would that dialogue be in political terms? It is fair enough to propose that peace parks would promote dialogue and the culture of peace between states. Nonetheless, questions arise as to why agreements over the use of natural resources such as water have not achieved that goal. The question is pertinent because agreements over peace parks in Africa are a recent phenomenon that should be assessed against a dense history of bilateral and multilateral agreements that range from trade to shared regional responsibilities in matters such as defence, for instance. South Africa alone has signed approximately 50 agreements over cross-border water issues since the 1920s. The most prominent of these is the Lesotho Highlands Water Project whose treaty was signed between the apartheid state and the Kingdom of Lesotho in 1988, and that has been inherited by post-apartheid South Africa. Similarly, Angola and Namibia have reaffirmed the colonial agreements on the Cunene River in 1990 and also incorporated Botswana to establish the Okavango River Basin Water Commission in 1994 (Pinheiro *et al.* 2003). These examples have been used here to demonstrate that African states have been involved in cross-border natural resources long before the advent of peace parks, and to argue that the prospects for peace through ecology should be evaluated against the experiences of other cross-border projects, such as shared water resources. The argument is valid in that water scarcity has generally been treated as a threat to peace among riparian states. The question of why existing cross-border water projects have not been used as the foundation of peace parks should be answered if we are to arrive at a useful meaning of peace in peace parks.

It follows that the peace–environment nexus remains problematic in Africa as elsewhere. In Africa, different avenues have been explored to validate that nexus. With regard to transboundary water resources, international water basins have recently been incorporated into the language of peace parks. For example, the Okavango River basin is one of the basins that Green Cross International has included in its Water for Peace Programme. Accordingly, Turton (2003) has invoked Ohlsson's concepts of negative and positive peace to argue that SADC should use its transboundary water resources towards fostering political stability. Significantly, the staunch supporter of the peace parks concept, the Washington-based Biodiversity Support Program, has referred to major water basins in southern Africa to support the peace parks argument against the division of ecosystems by international boundaries. In other words, transboundary water basins are increasingly been used to promote the idea of peace parks, though peace parks are more about wildlife and tourism than the protection of water basins. The proposed Kavango–Zambezi (KAZA) TFCA between Angola, Botswana, Namibia, Zambia and Zimbabwe is a case in point. The main aim of the KAZA TFCA is to establish the elephant corridor for purposes of ecotourism.[36] This is clear from Braack's

(2006, 9) comments that, 'perhaps the major contribution CI is making in assisting the five governments achieve their ambitious vision is by implementing a broad programme aimed at enabling elephants and wildlife to move freely across political boundaries over a strategically expanded and interconnected series of protected areas across the five countries.' Read from this perspective, the idea of peace parks is rooted in various sources that can conveniently be used at different stages in the development of peace parks.

Luckily for peace parks in Africa, the emergence of Wangari Maathai as the Nobel Peace Prize winner in 2004 served to highlight the environment–peace nexus. The focus of Maathai's Green Belt Movement on tree-planting for ecological and socio-economic purposes became useful in elevating trees as a symbol of peace – supposedly in line with African traditions[37] (*Earthyear* 2005). Poverty, more than any other life experience in Africa, has effectively been used to demonstrate the peace–environment nexus. In the view of proponents of peace parks, environmental threats are mainly caused by local and international competition over the wild areas of the continent, development projects and the condition of the African state. The poverty of local populations is considered a threat to the environment mainly because it is widely held that there is a causal link between poverty and environmental degradation.[38] Unsurprisingly, the staunch supporter of peace parks, Anton Rupert, has described unemployment in southern Africa as a time bomb, which can only be defused by employment that would be generated by the creation of peace parks (PPF 1997). Rupert described the threat to Africa's biological diversity for which he proposed the solution in these words:

> The basis of the problem is widespread poverty, caused by massive unemployment and a chronic lack of skills and entrepreneurial training, often leaving poor people living in overcrowded areas with little alternative but to destroy the very resource base on which their survival depends … It may seem surprising that the time has never been better for setting up a network of peace parks in Africa to address the concerns of rising unemployment and poverty, the accelerating loss of biological biodiversity, and the inveterate civil unrest across international boundaries (PPF 1998, 5).

In other words, it is believed that diverting the livelihoods of local populations towards earning income is the most appropriate strategy to avoid conflict over natural resources. A full discussion of how this operates on the ground is given in Chapter 6. For the moment, our attention is on the framing of environmental threats as part of strategies for marketing peace parks. There is a view among proponents of peace parks that dwindling world resources would trigger international competition over resources, and Africa would be an obvious target as a result of its abundant natural resources.[39] Resulting from this view is the need for peace parks as a measure for protecting the continent's biodiversity. Ironically, international players have been involved in the promotion of peace parks in southern Africa (see Chapter 4), raising the question of who is protecting biodiversity for, or against, whom?

Proponents of peace parks have carefully defined the threat posed by the African state to biodiversity in ways that would not alienate the state, since the state is required to formalise the creation of peace parks. There is therefore a tendency among peace parks ideologues to refrain from blaming post-independence African states for Africa's biodiversity loss or mismanagement. Instead, they prefer to highlight general issues such as lack of funds and capacity, weak legislative frameworks for managing protected areas, and so on (Peace Parks

Foundation 1998; Tukahirwa 2002). The silence on the weaknesses of the African states and their leaders with regard to biodiversity loss is in sharp contrast to the environmentalists' strongly held view that governments all over the world have abused their powers to the detriment of the environment and that states have lost their power to enforce environmental laws against powerful multinational companies (Brubaker 1995; Conca 1995). I insist that the failure by proponents of peace parks to blame post-independence African states and their leaders is a strategic move towards galvanising support from African leaders. Unsurprisingly, leaders in SADC have been made patrons of the Peace Parks Foundation,[40] and the peace parks agenda has become that of NEPAD and the African Union (AU), as we shall see below.

NEPAD, the AU and peace parks

Two important continental initiatives, namely, the launching of NEPAD in 2001 and the formation of the AU in the same year, have been useful for the peace parks initiative. Firstly, they opened up the entire continent for all sorts of projects, including peace parks, which could be pursued without the arduous task of explaining the relevance of such projects to a highly sensitive political environment. Secondly, the aims and objectives of NEPAD and the AU are too broad to allow internal and external interested parties to initiate or promote their projects in Africa under the banner of realising the aims and objectives of NEPAD and the AU. The primary objectives of NEPAD are stated as: (i) to eradicate poverty; (ii) to place African countries, both individually and collectively, on a path of sustainable growth and development; (iii) to halt the marginalisation of Africa in the globalisation process and enhance its full and beneficial integration into the global economy; and (iv) to accelerate the empowerment of women (NEPAD 2003). To this end, a healthy and productive environment is seen as a prerequisite for NEPAD (NEPAD 2001).

The direct link between NEPAD and peace parks emerged in the form of the Environment Initiative, which targeted eight sub-themes for priority interventions. These include the sub-theme[41] on Cross-border Conservation Areas, which 'seeks to build on the emerging initiatives, seeking partnerships across countries to boost conservation and tourism and thus create jobs' (NEPAD 2001, 35). The fact that one of the co-founders of NEPAD, President Wade (Senegal) has been the Co-ordinator of the Environment Initiative in the Heads of State Implementation Committee of NEPAD and that African Ministers of Environment are responsible for environmental matters in NEPAD, accounts for the high profile of the Environment Initiative within NEPAD. Indeed, the Initial Action Plan for NEPAD included the environment as one of the six Sectoral Priorities,[42] thus paving the way for the development of NEPAD's Environmental Action Plan in 2002. The Environmental Action Plan covers: land degradation; drought and desertification; coastal and marine resources; freshwater resources; better management of habitats, species and genetic resources; alien invasive species; management of non-renewable resources and impact of urbanisation; and HIV/AIDS and war/population dynamics (NEPAD 2002a). Unlike other priority areas, NEPAD's Initial Action Plan on the Environment did not provide clarity on how the focus areas on the environment would be handled. The United Nations Environment Programme (UNEP) and GEF facilitated the development of NEPAD's Action Plan for the Environment Initiative that was officially launched in Algiers on 15–16 December 2003.[43]

Five of the 12 specific objectives of the Action Plan for the Environment Initiative that have a direct bearing on peace parks are:

- To promote the sustainable use of Africa's natural resources and strengthen public and political support to subregional and regional environmental initiatives;
- To support the implementation by African countries of their commitments under the global and regional environmental conventions and other legal instruments to which they are party;
- To promote the integration of environmental issues into poverty reduction strategies;
- To foster regional and subregional co-operation to address environmental challenges; and
- To improve the institutional framework for regional environmental governance (NEPAD 2003).

Programme Area 6 of the Environment Initiative is devoted to transboundary conservation or management of natural resources. Arguably, existing conservation initiatives have largely shaped the Programme Area. For example, it is acknowledged that 'sustainable natural resource management requires co-ordinated transboundary policy and action in conformity with *existing agreements* among the countries concerned' and that 'a transboundary approach to sustainable use and conservation of natural resources within the Environment Initiative of NEPAD should be seen as a complement and extension of *existing national initiatives* and should build on these national level initiatives where the opportunity exists' (NEPAD 2003, 71). Using the typical language of supporters of peace parks, the Programme Area describes the economic and conservation benefits of transboundary collaboration as 'greater than would be achieved by countries working alone' (NEPAD 2003, 71).

It should be noted that NEPAD's Action Plan for infrastructural development also boosted the idea of peace parks, because the overall objective of the Plan is to develop 'trade corridors without barriers'[44] so as to minimise transit times and other transit costs[45] along Africa's main trading routes (NEPAD 2002b). In practice, this means doing away with double inspection and clearance of traffic by two sets of authorities at a particular border. Subsequently, southern African countries have embarked on transforming the border operations from two to a one-stop operation, and the Trans-Kalahari Corridor is one of the pilot projects. NEPAD assists in mobilising funds and securing political action with regard to concluding bilateral and multilateral agreements. Arguably, the move towards a one-stop operation at border posts augurs well for the idea that tourists entering peace parks should not be subjected to multiple immigration controls. In this sense, transforming immigration controls in peace parks appears to be a practical implementation of NEPAD.

Obviously, the launching of the Action Plan for the Environment Initiative was a culmination of events and negotiations. Indeed, relevant ministers have inserted peace parks in the NEPAD agendas. According to the International Co-ordinator of the Great Limpopo Transfrontier Park, George Ferrão, case studies of peace parks in southern Africa were presented to the African Union Summit that was held in Maputo, Mozambique, in July 2003 (Ferrão 2003)[46] ahead of the 5th IUCN World Parks Congress.[47] Arguably, environmentalists used the Summit to push for the endorsement of the peace parks initiative by African governments, the principal goal being to present peace parks at the World Parks Congress as one of Africa's strategies to conserve biodiversity. In essence, it was an attempt to develop a common vision of peace parks among African governments and to present that vision to the Summit as such. At the time, there was already a proposal to develop 22 TFCA sites, covering approximately 120 million ha, stretching from South Africa to the western Congo basin (Hall-Martin and Modise 2002). Whereas most southern African governments were already familiar with the activities of the

Peace Parks Foundation, governments in other regions in the continent had not yet endorsed the peace parks idea. Against this backdrop, the World Parks Congress provided environmentalists with the opportunity to market peace parks to the rest of the continent. Significantly, environmentalists hoped to use the Congress to persuade the members of SADC to adopt the proposed 22 TFCA sites as SADC's conservation agenda.

An important outcome of these developments was that the Second Ordinary Meeting of the Assembly of the African Union adopted the Action Plan on the Environment Initiative as per Assembly/AU/Decision 10(II). The Assembly endorsed the Action Plan on the basis that it articulates, among other things, the Plan of Implementation adopted by the World Summit on Sustainable Development in 2002.

Subsequent to the African Union Summit of July 2003 and the World Parks Congress that was held 2 months thereafter, the NEPAD Action Plan for the Environment Initiative was officially launched in Algiers on 15–16 December 2003. The proceedings of the occasion included the adoption of that Action Plan by the African governments in the form of the Algiers Declaration. The Declaration recommitted African governments to: (i) meeting the special needs of Africa as per Chapter 8 of the Johannesburg Plan of Implementation adopted by the World Summit on Sustainable Development; and (ii) entering into a structured dialogue by which roles and responsibilities among stakeholders will be identified at various levels.

The Declaration was historic in the context of the past and the present responses by African states to environmental issues. With regard to the past, the African Convention on the Conservation of Nature and Natural Resources was adopted in 1968 in Algiers at the height of Africa's liberation struggles. In contemporary Africa, the spirit of those struggles and the quest for the ownership of Africa's resources by Africans themselves is presented as the purpose of NEPAD, and the Algiers Declaration has been presented as the implementation of NEPAD. The Chairperson of the NEPAD Steering Committee, Wiseman Nkuhlu, linked NEPAD and the Algiers Declaration in these words: 'It is the first time in history that African leaders and their partners have converged in an African country [i.e. Algeria], in the context of NEPAD, to discuss strategies for addressing Africa's environmental and sustainable development challenges on such a comprehensive scale'.[48]

Other steps towards consolidating the Environment Initiative into the agenda of NEPAD included the Sirte Declaration on the Environment and Development (June 2004), the Dakar recommendations of the Conference of Ministers on resource mobilisation (October 2004) and the Dakar Declaration for the Enhanced Partnerships in the Implementation of the Action Plan for the Environment Initiative of NEPAD (2005). The Dakar Declaration not only appealed for international support for the Environment Initiative, but also urged African governments to commit themselves to the Initiative by integrating it into their development plans. More importantly, it adopted the proposal for the creation of the African Environment Facility[49] at the African Development Bank. As the Declaration emphasises the implementation of the Action Plan of the Environment Initiative through existing regional blocs, southern African states have the opportunity to implement peace parks as part of the NEPAD plan of action.

Clearly, the adoption of NEPAD as a programme of the AU in 2001 means that NEPAD and the AU are inseparable. With regard to peace parks, the following objectives of the African Union have also been used to support the creation of peace parks: (i) to achieve greater unity and solidarity between the African countries and the peoples of Africa; (ii) to defend the sovereignty, territorial integrity and independence of its Member States; (iii) to accelerate the political and socio-economic integration of the continent; (iv) to promote peace, security, and stability on the continent; (v) to promote sustainable development at the economic, social and cultural levels as well as the integration of African economies; and (vi) to co-ordinate and harmonise the policies between the existing and future Regional Economic Communities for the gradual attainment of the objectives of the Union. Both the objectives of NEPAD and the AU, particularly peace and security – as one of NEPAD's conditions[50] for sustainable development – and the principle of the African ownership of development projects chime with the political rhetoric of peace parks as a home-grown initiative for African unity, sustainable development and peace. Advocates for peace parks have presented the links between peace parks and NEPAD and the AU as follows:

> The political leadership of SADC, now involved in the even more ambitious development of the African Union and the New Partnership for Africa's Development (NEPAD) is committed to co-operation. Not only will TFCAs allow for the concrete manifestation of these ideals, they will also tap into the current world mood of assisting regional developments rather than unilateral ones (Hall-Martin and Modise 2002, 11).

> In Africa, the establishment of TFCAs is one of the projects that epitomises the implementation of the New Partnership for Africa's Development (NEPAD) (*International Tourism Monthly* 2003).

> President Thabo Mbeki in particular supports [the concept of peace parks] as it fits in with the New Partnership for Africa's Development (NEPAD) (PPF 2003, 1).

Politicians have also endorsed the linkages between NEPAD and the AU and peace parks. For instance, on the occasion of the signing of the Treaty of the Great Limpopo Transfrontier Park, South African President Mbeki commented that 'the birth of the Great Limpopo Transfrontier Park today, tells the citizens of our continent that the AU and NEPAD are not merely a set of good and grand ideas whose accomplishment will be in the distant future. This Transfrontier Park says that each passing day transforms the dream of an African Renaissance into reality' (Mbeki 2002). In the same vein, South Africa's former Minister of Environmental Affairs and Tourism, Valli Moosa, is of the view that South Africa has impacted on the rest of the African continent and on NEPAD through the establishment of transfrontier parks as evidenced by the fact that 'transfrontier parks are now mentioned in the NEPAD programme and are an integral part of the African development programme' (Moosa, cited in MacLeod 2004, 9).[51] Undoubtedly, peace parks have become the agenda of NEPAD and the AU. The penetration of peace parks into such continental structures has nevertheless been facilitated by the adaptation of global propositions of, and assumptions about, transfrontier parks to the African conditions. Assumptions about borders in peace parks can be used to substantiate this claim.

Border narratives and peace parks in Africa

> Ongoing conflicts in far too many parts of Africa, Asia, Europe and South America no longer impact only the military forces involved. Women, children and the aging make up the bulk of the victims, totalling literally millions of displaced people who have lost their homes, their dignity and self-respect. The human costs of

> these conflicts are immense, with the inevitable associated collapse or stagnation of economic development, and the acceleration of environmental degradation. Nowhere is this better encapsulated than on the borders of Uganda, Rwanda and the Democratic Republic of the Congo, where well over one million displaced people live in a twilight zone of fear of attack and harassment, in the midst of disease, malnutrition and a collapsed infrastructure (Peace Parks Foundation 1999, 1).

That the borders of Uganda, Rwanda and the Democratic Republic of the Congo are inhabited by people fleeing from conflict and genocide is indisputable. However, reference to the borders of the three countries as a site of human suffering serves to draw attention to Africa's colonial borders that should be changed in order to promote peace in Africa. Indeed, protagonists of peace parks often refer to Africa's borders as colonially imposed, irrational, and as a cause of political instability; problems that could effectively be resolved through the establishment of cross-border peace parks. In so doing, Africa's borders become the vortex around which a common anti-colonial language is developed and shared among politicians, ecologists, pan-Africanists, consultants, the tourism industry and border communities, as revealed by the following statements:

> The fences erected by *erstwhile colonialism* would be brought down on systematic basis to create a common conservation area that straddles three countries (Moosa, cited in Dube 2001, 28).[52]

> The new park [i.e. GLTP] will boost regional co-operation between the three countries, promote peace and security in the region, and help deal with illegal animal poaching. It will restore the integrity of an ecosystem artificially segmented by *colonial boundaries*, opening up the natural migratory routes of African elephants, endangered species such as the roan antelope, and other animals (South Africa 2003).

> Throughout the 1950s, Pan Africanism denounced colonial boundaries and called for their dissolution to facilitate unity among African people … Pan Africanism therefore has an unfinished business with our arbitrary borders which must be addressed (Gakwandi 1996, 185, 187).

> One has to recall that the Berlin Treaty[53] of 1884 dealt with African territories like a pack of cards. National boundaries then proclaimed cut across tribal and clan groupings as well as wildlife migration routes, fragmented ecosystems and threatening biodiversity. The establishment of peace parks strives to correct these past injustices and ensure that a high level of biodiversity is maintained through the joint management of these resources (www.peaceparks.org).

Four observations can be made from the quotations above, the first being to blame colonialism for imposing international borders on Africa. Clearly, colonial borders were not imposed on Africa alone and Africa was not the only continent that had been colonised. The emphasis on the act of imposing boundaries on Africa is meant to gain political support. In former colonies that have become part of the developed world the idea of decolonising boundaries is almost absent. It should be pointed out that colonial borders in Africa were not limited to national borders but also appeared as internal administrative units. There were also other internal physical boundaries that developed around properties such as farms, which were effective in promoting racial planning (e.g. in Namibia, South Africa and Zimbabwe) and the colonial policy and strategy of divide-and-rule. This begs the question of why colonial borders that had been imposed within African states are not referred to in the promotion of peace parks. In order to answer this question we need to understand the proposition on boundaries discussed above. Any concern with internal colonial boundaries would be counterproductive to the intended goal of creating international peace parks in Africa, and would force proponents of peace parks to come face to face with the complex realities of the politics of the African state. Boundaries around properties such as farms, are not part of the main border narrative for peace parks, because reference to the colonial history of such boundaries in Africa has the potential to disrupt the creation of most peace parks, particularly where land dispossession in the continent still has to be resolved. In fact, supporters of peace parks have been concerned that land reform in southern Africa would disrupt the peace parks initiatives (Ramutsindela 2002, 2004b). The main point here is that Africa's international boundaries were singled out for the promotion of peace parks, because the aim is to create peace parks between and among states rather than within particular states. In the process, the nature and histories of Africa's international borders were generalised in order to promote a hegemonic perspective of the border and its problem. This brings us to the second observation: the obsession with fences as Africa's boundary markers.

Africa's borders are presented as if they are all marked by fences or objects that should be removed. This has clearly been captured by Michler's discussion of the roles of the Peace Parks Foundation as to 'bring down the fences' (2003, 79). The picturing of African boundaries as marked by fences also appears in official documents. In the conceptual plan for the establishment of the Great Limpopo Transfrontier Park, African boundaries were described as follows:

> Political boundaries were drawn with very little regard for their ecological consequences. National borders were sketched on a map to suit strategic defence needs, mineral riches and natural resource needs, demographic and other realities of the time. These political boundaries later became international fences, very often severing traditional migration routes of animals or otherwise impending natural processes (South Africa 2000, 1).

Evidently, international boundaries in Africa are not all marked by fences. This is also the case in southern Africa where the seed of peace parks was first planted in the continent. For example, the Gariep[54] River marks the boundary between Namibia and South Africa, while cemented blocks mark that between Botswana and South Africa. That is to say that there are no fences marking the international boundaries between these countries. In some cases the fence exists but is not effective as is clearly the experience on the Lesotho–South African border where locals have long cut down the fence – resulting in people, livestock and wildlife crossing the boundary between the two countries with little control. The focus on fenced boundaries diverts public attention away from the fact that conservationists are equally responsible for erecting most of the fences around protected areas. Conceptually, in the language of peace parks, the fence becomes a metaphor by which certain fences are denounced as obstructive to the 'public good'. In Africa, that metaphor is constructed in the contexts of a selected history of colonialism and the on-going initiatives for African renewal.

The third observation about the framing of Africa's borders by advocates of peace parks is that the emphasis on the negative effects of Africa's colonial borders has been on wildlife. The Botswana-based Kalahari Conservation Society commented that 'the African boundaries have since establishment, been an ecological problem as they did not only separate communities, but have cut across traditional migratory routes important to various wildlife species' (Tema 2000, 1). I argue that reference to the view that colonial boundaries separated African communities is made to achieve political correctness, because, unlike in the case of ecosystems where there are clear strategies to reunite areas belonging to the same ecosystem, official and non-official documents on peace parks are silent on how communities would be reunited across political borders. On the ground, communities are simply encouraged to communicate across political borders in the form of exchange visits. For instance, officials administering the Great Limpopo Transfrontier Park organised an exchange visit among the Tsonga-speaking communities from Sengwe and Chiredzi (Zimbabwe), Shingwedzi and Limpopo (Mozambique) and Makuleke (South Africa) in 2003 (Ferrão 2003). Similar visits

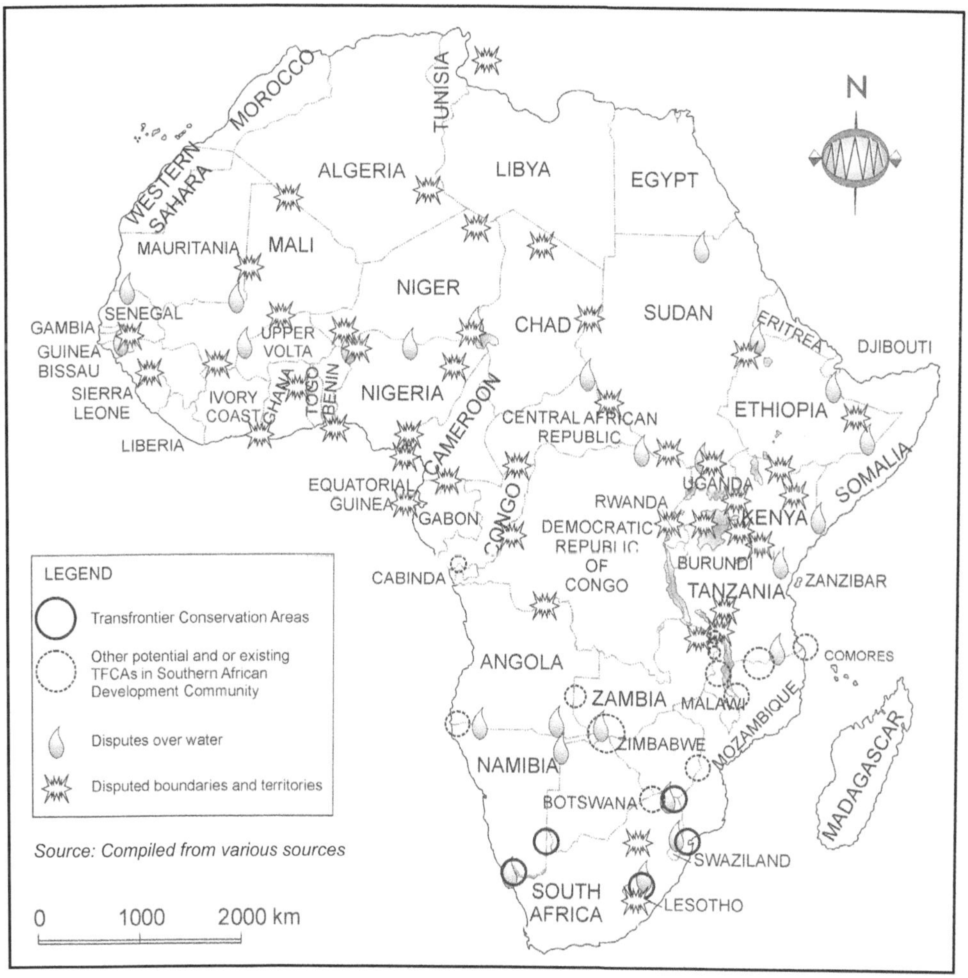

Figure 3.1 The location of disputed borders and TFCAs.

have taken place between the Nama in South Africa and their counterparts in Namibia as part of the promotion of the Ais-Ais/Richtersveld Transfrontier Park. In the Maloti–Drakensberg TFCA the visits between communities in the Upper uThukela Valley (South Africa) and the Senqu Sources area (Lesotho), and the BaTlokwa leaders in the former Qwa Qwa bantustan and Lesotho were arranged in early 2006 (Maloti–Drakensberg Transfrontier Project 2006).

The last and fourth observation is that advocates for peace parks in Africa promote a narrow view of Africa's boundary problems in order to propose a common solution – the creation of peace parks. Boundary disputes in Africa fall into different categories involving territorial claims, the functions of boundaries, boundary positions and claims over resources. In contrast to the populist view that peace parks are instruments for peace in disputed borders, there is no congruence between the location of peace parks and disputed borders in Africa (Figure 3.1).

In summary, the Waterton/Glacier Peace Park experience has become a model for peace parks worldwide, including in Africa. Though the model is underpinned by the rationale for ecological integrity and peace, its focus is on the international scale. The result has been the dominance of inter-state rather than intra-state peace parks, even in places such as Africa where internal factors contribute significantly to political instability. In Africa, proponents of peace parks have used boundary narratives that are relevant to ongoing discussions about the future of Africa and the unity of its people. Those narratives have specifically been used to integrate peace parks to the long-established political ambition of African unity. Most commentators on African unity have ascribed the lack of that unity to the divisive nature of African boundaries, hence repeated calls have been made that African boundaries should be redrawn (Bello 1995; Gakwandi 1996) to achieve political unity or be adjusted for the African renaissance (Breytenbach 1999). In both instances, the concern is with international boundaries, which are a target for peace parks.

Endnotes

1 This was a response by former President Mandela to the journalist Peter Godwin's question about what made Mandela decide to support the peace parks plan (Godwin 2001).

2 The foundation for collaboration was laid by the personal relationships among staff of both parks.

3 www.peace.ca/rotarypeaceparks.html (accessed 24/5/2005).

4 The main reason given for this failure is that the former US President Franklin Roosevelt's wish to establish the international park on the US–Mexico border was interrupted by the outbreak of World War II. In contrast to the establishment of the first transboundary peace park by the United States, the situation on the US–Mexico border was marked by the fortification of the border on the side of the USA. Recent analyses show that the United States embarked on the enforcement of the US–Mexico border to control immigration and to pursue economic interests (Andreas 2000; Nevins 2002; Purcell and Nevins 2005). The attempt to develop an international peace park between the two countries became public in the 1990s, though it is said that the first proposal for that park was made in the 1930s.

5 It is considered the First Rotary Peace City of the World (http://rosella.apana.au) (accessed 5/1/2006).

6 The park was established in 2000 to mark the 40th anniversary of the death of the United Nations Secretary General, Dag Hammarskjöld, who died in a plane crash in Ndola (Zambia).

7 The Freedom Park was established in terms of the National Heritage Resources Act 25 of 1999.

[8] These are often implicit rather than explicit.

[9] The certification process was launched on 10 May 2003 under the theme 'Transboundary Parks – Following Nature's Design' (EUROPARC 2003).

[10] The same year that was declared the United Nations International Year of Peace. The tourism industry therefore capitalised on this declaration to launch the IIPT.

[11] A heated debate about the need to clarify the notion of a boundary appeared in the form of e-mails that were exchanged during preparations for the World Parks Congress held in Durban in 2003. The intention of those involved in the debate was to arrive at a common terminology that was to be adopted at the Congress.

[12] A detailed reference to John Hanks is given in Chapter 4.

[13] Leo Braack was the first international co-ordinator for the much-publicised Great Limpopo Transfrontier Park.

[14] As I will show below, the emphasis in Africa is on transcending colonial boundaries.

[15] www.missoulian.com/articles/2005/09/19/news/mtregional (accessed 9/1/2006).

[16] They may not require the same solution, either.

[17] Hamilton *et al.* (cited in Fall 2005) have suggested that protected areas that share common borders share common problems. A distinction should be made between sharing common problems and problems such as air pollution that spill over state borders.

[18] From the environmental point of view, the creation of Condor–Kutuku conservation corridor in 2004 is said to have contributed to the promotion of peace between the two countries.

[19] The TerraGreen website suggests that the Revolutionary Armed Forces of Columbia (FARC) has struck a peace deal with the Colombian government by agreeing to co-manage large swathes of protected areas that had been under the control of FARC (www.teriin.org/terragreen) (accessed 9/1/2006).

[20] The American Peace Society was created in 1828 through the merger of 36 societies from different areas (Carter 1992).

[21] The spirit of early Christianity was that waging war was incompatible with respecting 'that of God in every man' (Carter 1992, 2).

[22] The post-Cold War wave of ethnic revivalist movements in Central and Eastern Europe could be used to support this view.

[23] Incidentally, this happened at the same time when ideas of peace parks gained momentum.

[24] In promoting the idea of environmental security, Butts (1994) ascribed the political difficulties of Haiti to severe environmental degradation that eroded the resource base upon which the government should be established.

[25] Brock (1991) cautioned that the concept of environmental security may be invoked to defend the status quo of the world ecological order, hence the concept of sustainable development might appear appropriate.

[26] A Special Issue of *Geopolitics* was published in 2004 under the theme of 'the geopolitics of resource wars'.

[27] As far as I have been able to verify, the word 'peace' has not been recorded in treaties and memoranda establishing TFCAs in southern Africa.

[28] The policy expressed the objectives of South Africa's Total Strategy by which the apartheid state aimed to control the region for security purposes and to prevent the spread of communism to southern Africa.

[29] South Africa's withdrawal of its nuclear weapon programme has been used as a model for nuclear disarmament, which countries such as Iran, whose nuclear programme is condemned mainly by the West, should follow (Fabricius 2006).

30 Cawthra (1997) has suggested that the new approach to security in developing countries resulted in part from Structural Adjustment Programmes, which require shifting resources away from military security towards social and economic development.

31 The other threats were ethno-political conflict, small arms proliferation, the displacement of people, hunger, poverty, crime and corruption.

32 Private military firms operated in 11 African countries between 1995 and 2005 (Smith 2005).

33 See various chapters in this volume.

34 There could be variations in the tactics of war for biodiversity in national and transfrontier parks.

35 The Revolutionary United Front of Sierra Leone committed atrocities against civilians in its quest to gain control of the state, following its manifesto of 1991 (Savage and Rahall 2003).

36 In fact, the KAZA TFCA has been described as Conservation International's mega-elephant project (Braack 2006).

37 In seeking to establish the relationship between peace and the environment, reference has been made to the Kikuyu tradition by which elders carried branches from the *thigi* tree that 'when placed between two disputing sides, caused them to stop fighting and seeking reconciliation' (*Earthyear* 2005, 14). In contrast to the *thigi*, the branches from other trees can equally be malignant. *Phaladzane* (something that scatters) has been used among Venda-speaking people to sow discord among family members.

38 See Mabogunje (1984), Broad (1994) and Fairhead and Leach (1996) for the debate on this link.

39 Interview, Noel de Villiers, 20/9/2005, Kirstenbosch.

40 See chapters 4 and 5 for further details.

41 The other sub-themes are: combating desertification; wetlands conservation; invasive alien species; coastal management; global warming; environmental governance; and financing.

42 Five of these are education, health, regional infrastructure, agriculture, and market access.

43 This is known as the Algiers Declaration by which African leaders recommitted themselves to the special needs of Africa as recognised in Chapter 8 on sustainable development of the Johannesburg Plan of Implementation adopted by the World Summit on Sustainable Development in 2002.

44 These corridors include the: Northern Corridor (linking the Great Lakes countries with Mombasa); Trans-Kalahari Corridor (linking Walvis Bay in Namibia with Botswana and South Africa); Maputo Corridor (linking Maputo with South Africa, Swaziland and Zimbabwe); and Beira Corridor (linking Beira with Malawi and Zimbabwe).

45 It has been estimated that the costs of transit delays in southern Africa amount to US$48–60 million per year (NEPAD 2002b).

46 It has been reported that President Wade of Senegal was so impressed by the Great Limpopo Transfrontier Park project that he requested the information about the project to be made available to his own country.

47 The Congress was held in Durban between 8 and 17 September 2003.

48 www.un.org/special-rep/ohrlls/News_flash2004 (accessed 14/01/2006).

49 An equivalent of the Global Environment Facility of the World Bank.

50 Other conditions are: democracy and good political, economic and corporate governance; regional co-operation and integration; and capacity building.

51 It is reported to have listed transfrontier parks as one his most important achievements of tenure as Minister (MacLeod 2004).

52 The former South Africa's Minister of Environmental Affairs and Tourism.

53 See Katzellenbogen (1996) for a critique of Africa's boundaries and the Berlin Treaty.

54 Formerly known as the Orange River.

Chapter 4

Local initiatives and their regional and global connections

Transfrontier conservation areas (TFCAs) in Africa are established across state borders along the lines of the Waterton/Glacier International Peace Park model (see Chapter 3) while, at the same time, articulating the ecological concerns of the environmental movement (see Chapter 2). In the African continent, the seed of TFCAs was planted in southern Africa, where local nature conservation initiatives and institutions played a key role in both the development of TFCAs and in the shape that TFCAs were to take later on. The map of TFCAs in southern Africa has its roots in activities that occurred long before the formalisation of TFCAs. In this sense, understanding TFCAs demands that we should appreciate the historical conditions in which they developed. The purpose of this chapter is to sketch the historical background and activities of the Southern African Nature Foundation (SANF) in order to aid our understanding of the development of TFCAs in the region and the rest of the continent on the one hand, and to establish the links between TFCAs in Africa and the rest of the world, on the other hand.

The SANF: a frontrunner of the Peace Parks Foundation

Almost all analysts of TFCAs in Africa have referred to the PPF that was established in 1997 as the think tank behind the establishment of TFCAs in southern Africa. The role of the PPF in promoting TFCAs in the southern Africa region can neither be denied nor ignored. However, the idea of TFCAs in the region precedes the formation of the PPF. It could be argued that the PPF continued in the footsteps of its predecessor, SANF, which was founded by South African billionaire Anton Rupert, on 14 June 1968.[1] It has been suggested that Rupert started SANF at the request of Prince Bernard of the Netherlands, who was President of WWF International at the time (SANF 1988, 1993). In essence, the establishment of SANF was both a response to the failure of WWF International to open a branch in South Africa and the outcome of Rupert's interest in conservation projects in the region. According to Dommisse and Esterhuyse (2005) Prince Bernard wanted to establish a WWF branch in South Africa in 1966, but his criticism of the South African government's decision to subdivide the Etosha National Park in Namibia[2] led to Prime Minister Verwoerd's refusal to support the creation of WWF branch in South Africa. Following Verwoerd's death in 1966, the attempt to establish WWF–South Africa was frustrated by the Wildlife Society's unwillingness to become a branch of WWF, hence the formation of SANF.

The establishment of SANF as an affiliate of WWF in 1968 meant that South Africa was the first country in Africa to join the WWF family. The second African country to follow suit was Nigeria in 1989. In global terms, South Africa was one of the first ten countries to participate in WWF International. When WWF International's fundraising agency was established in 1971 to raise funds for operational and scientific costs, South Africa had 71 contributors, making it the fourth on the list of WWF donor countries (SANF 1993; Dommisse and Esterhuyse 2005).[3] Against this backdrop, the establishment of SANF and its relationship with WWF International conferred on South Africa a leadership role in WWF activities in Africa. Unlike other WWF affiliates, which operate within the boundaries of the countries in which they are established, SANF activities went beyond the borders of South Africa. It represented the activities of WWF in South Africa and its 12 neighbouring countries (Figure 4.1). It should be noted that SANF, like the apartheid state, recognised the nominally independent states of Transkei, Bophuthatswana, Venda and Ciskei (TBVC states), as South Africa's neighbours.

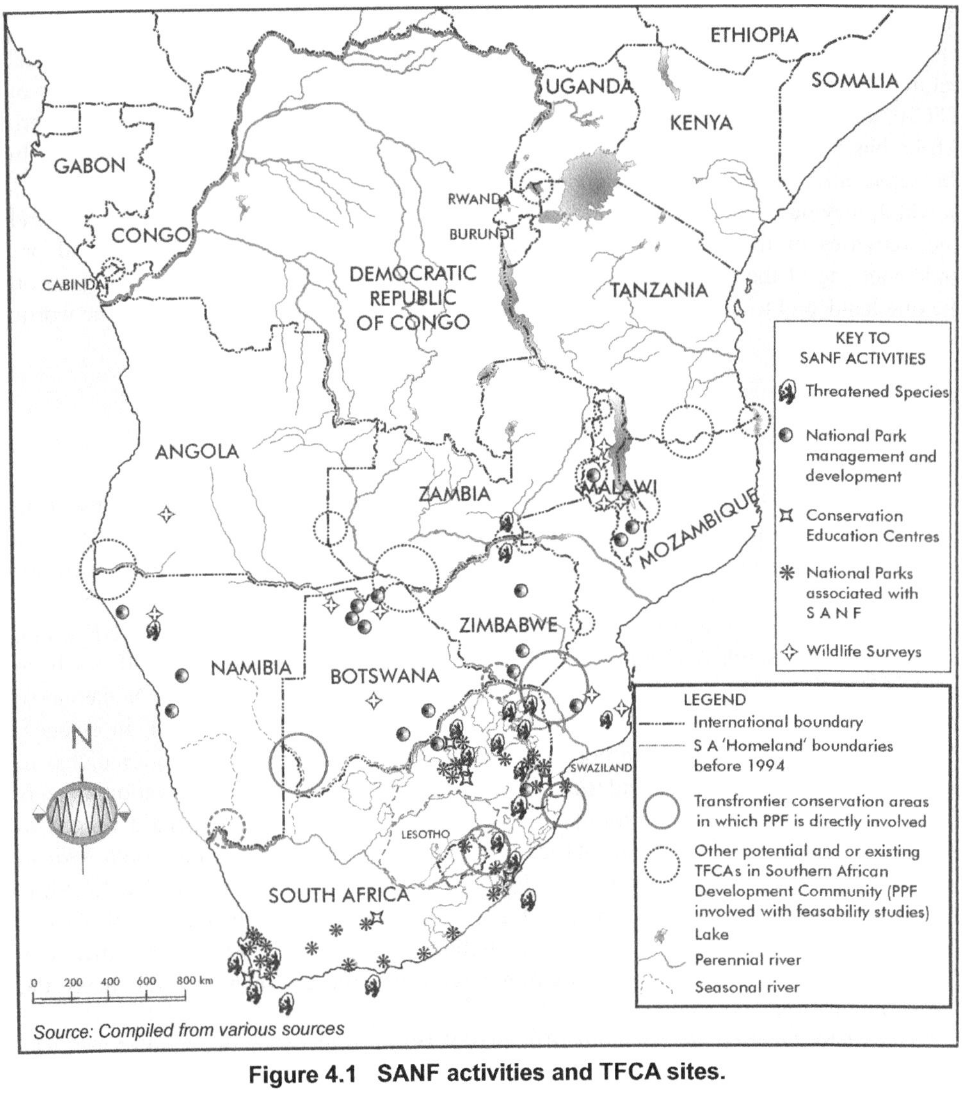

Source: Compiled from various sources

Figure 4.1 SANF activities and TFCA sites.

The geographical reach of SANF is relevant to our discussion for two main reasons. Firstly, it suggests a regional perspective of SANF that was later to be followed by the PPF. In no way do I imply that SANF propagated the establishment of TFCAs from the onset. Rather, the point is that SANF had its projects scattered throughout the region, because it viewed the promotion of conservation in the region as its mandate. As I will show later in this chapter, most of the SANF projects in the region formed the basis on which TFCAs were launched. Secondly, the sphere of influence of SANF also meant that the Foundation could connect with state and private conservation bodies in different parts of the region, thereby establishing a strong network of agencies.[4] In practice, conservation bodies in the region became instruments for SANF, more so because SANF was not a statutory body but operated as a non-governmental organisation. The aim of SANF was to conserve the natural heritage, with a strong bias towards wildlife. This is not surprising since SANF's international affiliate, WWF, was, from its inception in 1961, mainly concerned with the protection of wildlife.[5] This bias is clear in the description of the area under the responsibility of SANF as 'an area 9 times the size of France, containing amazing – and irreplaceable wealth of wildlife' (SANF 1984, 1). In fact SANF president Rupert was of the view that the whole western Transvaal[6] should be farmed with wildlife and that cultivation in that area should be regarded as a criminal activity (SANF 1988).[7] Accordingly, SANF established 19 national parks and reserves in 11 southern African countries in its first 15 years of existence. In the Foundation's view, conservation areas were to be carefully planned in order to promote a network of conserved land across the country to 'ensure the survival of [the] wildlife heritage while enabling people to benefit in countless ways from [the] tremendous indigenous natural wealth' (SANF 1985, 7). According to Rupert (SANF 1985), this was indeed the Foundation's major goal since its establishment in 1968. Against this backdrop, the vision for a network of conservation areas in the region was conceived during apartheid in South Africa. Strategies to achieve that vision included the launching of the South African Heritage Programme, the establishment of the National Parks Trust Fund and the re-establishment of wildlife riches through the translocation of wildlife.

The South African Heritage Programme was a partnership between the Department of Environmental Affairs, provincial conservation bodies and SANF,[8] which aimed to encourage individual landowners to register their land as heritage sites for conservation. Under the South African Heritage Programme, the landowner received a certificate signed by the State President[9] and a bronze plaque in return for the commitment to give at least 60 days warning of a major change in land use (SANF 1985). By 1985, almost 50 sites had been registered under the Programme. For its part, the National Parks Trust Fund was initiated by SANF for purposes of creating new national parks or expanding existing ones through the acquisition of land. To this end, the P.W. Botha government contributed R4 million to the Fund in 1984/85. This donation implies that the apartheid government supported the activities of SANF, raising the question of whether SANF financed the apartheid state.[10] The relationship between SANF and the state was further strengthened when the former Foreign Minister Pik Botha's Private Secretary, Frans Stroebel, joined SANF in 1980. By 1988, Stroebel had become the Director of SANF.

Arguably, land acquisition for conservation was also motivated by the belief that it would be almost impossible for any post-independence government in Africa to change land use in protected areas. As Rupert maintained, 'once land has been bought ... and handed over for conservation, the chances that any government will in later years use that ground for purposes other than conservation are remote. In Africa, none of the independent black nations has done this despite their political philosophies' (SANF 1988, 11). Conceptually, nature conservation can be used as an instrument for protecting land against the whims of political changes. The most instructive example of this is the switch from agriculture to game farming in anticipation of, or as a response to, land reform programmes, as has been the case in Botswana, Namibia, South Africa[11] and Zimbabwe. Essentially, game farming falls under the broad category of conservancies that are established as a mechanism for increasing the size and diversity of protected areas with minimal cost to the state. In the context of political changes, particularly land reform, game farming has been viewed as the recolonisation of the countryside (*Mail & Guardian* 2005).

The most defining character of SANF was that it was strongly linked to the business sector. Accordingly, Rupert commented that when the Foundation was formed in 1968, 'leaders in commerce and industry took up the challenge of conservation in southern Africa' (SANF 1984, 5). The Foundation's membership was restricted to companies, and corporate membership was graded according to annual membership fees (Table 4.1). In 1988, the Foundation's Director, Frans Stroebel, defended the restriction on corporate membership on the grounds that SANF did not want to compete with like-minded organisations such as the Wildlife Society, which relied on private individuals. This begs the question of the interests that companies had in nature conservation. A careful reading of the supplements that SANF published in the *Financial Mail* in four consecutive years between 1987 and 1990 reveals that SANF aimed to consolidate the convergence between business and conservation interests. The fact that these publications also appeared in the dying years of the apartheid state suggests that SANF sought to reposition itself in changing political environments. Such a repositioning required the redefinition of the roles of companies from that of simply promoting the establishment of protected areas to that of shaping southern Africa as a future market. In other words, political changes heightened anxieties in commercial and conservation circles alike, thus creating new avenues for stronger collaboration between business and SANF.

Table 4.1 Categories of the corporate membership of SANF.

Category of membership	Annual membership fee
Founder Members	R5000 and above
Sponsor Member	R2000 – R5000
Associate Member	R500 – R2000

Source: SANF 1987

With regard to nature conservation, more funds were required to finance projects that were meant to rescue wildlife from the risks of political changes in the region. This was clearly captured by the Foundation's annual reports of 1988 and 1990, which had pictures of animals rescued by huge boats as in the biblical Noah's Ark. That is to say that companies were encouraged to pay for the high cost of animal translocation. For example in 1987, the Foundation assisted in the transfer of the black rhino from the Zambezi Valley to Hwange National Park in Zimbabwe by means of a helicopter. And, 'as a *quid pro quo* for its assistance the Foundation managed to negotiate a deal through which a breeding herd of six of these black rhino were shipped from the Zambezi Valley to Swaziland where they were extinct in return for the transfer of six white rhino from Swaziland to Zimbabwe where there [were] none' (SANF 1987, 10). In the same deal, 18 elephants were transferred from the Kruger National Park to nature reserves in Swaziland. The translocation of these animals followed on the experiment that had been carried out during the creation of Pilanesburg National Park in the former Bophuthatswana bantustan in 1979. Dubbed 'Operation Genesis', some 4000 animals from Botswana, Namibia, Natal and the Cape were brought to the park in 1979/80. In the Foundation's view, the translocation in Pilanesburg was not only one of the biggest game translocation operations in history but it was also 'designed as a blueprint for the creation of new nature reserves in developing countries' (SANF 1985, 26).

Between 1987 and 1990 the Foundation also launched a campaign to highlight the benefits that would accrue to companies that participated in its activities. SANF's manager for public relations and marketing, Carine van der Merwe (SANF 1989), promoted the view that companies should move away from funding conservation from 'the corporate charity chest'[12] to embracing conservation as a marketing tool for profit-making. In that way, both conservation and the private sector would benefit. The main benefits at the time were that:

(a) Companies linked to SANF will generate international interest in their business through the world's largest private conservation organisation, WWF. SANF emphasised the point that its sister body, WWF–Switzerland, had more members than any one political party; Canada had one person in five actively involved in wildlife-related activities and both the Kruger National Park and the Kirstenbosch Botanic Garden in South Africa received about half a million visitors each per year. These examples were used to demonstrate the potential market for companies sponsoring conservation.

(b) Companies could make environmental sponsorship part of their social responsibility. The SANF promoted the view that 'a conservation sponsorship enables the company to highlight its role in the community and reinforce its corporate personality' (SANF 1987, 3).

(c) Conservation projects could be tailor-made to suit the company's marketing needs and improve its image. With regard to the market, companies would save advertising costs while expanding the market. This was to be achieved through SANF's educational newspaper, *Our Living World*, which had an estimated circulation of half a million and a potential circulation of more than one million, making it (in SANF's view) 'the biggest circulation newspaper in Africa' (SANF 1987, 6). In this way, SANF offered to use its media to publicise companies that supported its conservation projects.

(d) SANF mooted the idea of the 'loans for nature scheme' in which companies could contribute capital on the understanding that a refund for their money was guaranteed after 5 years.

Companies responded by developing brands that emphasised a special relationship with SANF, and also participated in activities that exposed them to a large market. SANF sold itself to companies as the leading conservation organisation in Africa, and 'by far the heavyweight in local non-governmental organisations' (SANF 1988, 8). In 1988 it had a budget of R10 million, which placed it at the top of the list among other environmental NGOs operating in the country at the time. It achieved its 'dominant financial status because of its links to the WWF' (SANF 1988, 8). Its dominance did not only attract companies, but was also reinforced by those companies that were associated with SANF. For example, the spokesperson for South African Airways (SAA), Marie-Heléne Maguire, commented that, 'we investigated all the local conservation bodies and decided that the SANF was a leader in its field' (SANF 1990, 50). The SAA hoped to benefit from the activities of SANF in transporting tourists, who, according to the airline, came to South Africa because of the lure of the wild – hence the wild had to be protected. For its part, Cadbury came with a wildlife series of chocolate bars with wrappers that showed the riverine rabbit, cheetah, bat-eared fox and vervet monkey (SANF 1989) to demonstrate its commitment to nature conservation. AECI Explosives advertised itself as environmentally-friendly blasting consultants in the SANF pages in these words:

> Coleford is one of Natal's most picturesque nature reserves … it's the home of the rainbow trout. The Parks Board … had a problem. The Coleford stretch of the river consisted of a series of deep pools, linked by shallow areas, where the river ran over solid rock. In winter, especially, when the flow is at its lowest, these shallow stretches were unable to support any significant number of trout. The obvious answer was to deepen the shallow areas. The question was, how? And how to do it without upsetting or harming the resident trout? The Parks Board contacted AECI Explosives and Chemicals Limited, and one of our explosives engineers, Tony Goetzsche, immediately went down to Coleford … and recommended deepening three shallow areas using our explosives … 75 kilograms of explosives were used, in five separate blasts. They were all highly successful and the river flowed freely – deep and strong … So if you would like to make your blasting operations safer and more productive … call one of our expert engineers or blasting consultants (SANF 1988, n.p.).

What is instructive about the AECI's advert above is the attempt by companies to link their activities with nature conservation. The deepening of the river by AECI in Coleford did not only eliminate the unwanted[13] 'gillieminkies' or mosquito fish in that river, but also created conditions for the population explosion of the trout, leading to Coleford becoming[14] a trout fisherman's paradise.

In short, the main focus of the campaign was to promote conservation sponsorships as a marketing strategy. Corporate members were persuaded to shift their focus from sports sponsorships, which were considered to have reached market saturation. Sport was also at risk of boycotts at the time – the height of the struggle against apartheid. SANF's campaign to increase sponsorship paid off, because corporate membership increased dramatically in 1990 (Figure 4.2), the year in which SANF had its highest income of R32 million in its 25-year history. By 1993, it had about 320 members, far above the initial figure of 72 recorded in 1968. The positive responses of companies towards conservation in the last decade of the apartheid state should be understood in the context of South Africa's economy at that time. The economy was weak and, following President Botha's refusal to effect significant political changes, there was 'a huge haemorrhaging of capital from South Africa' (Hirsch 2005, 24). Under these conditions SANF encouraged

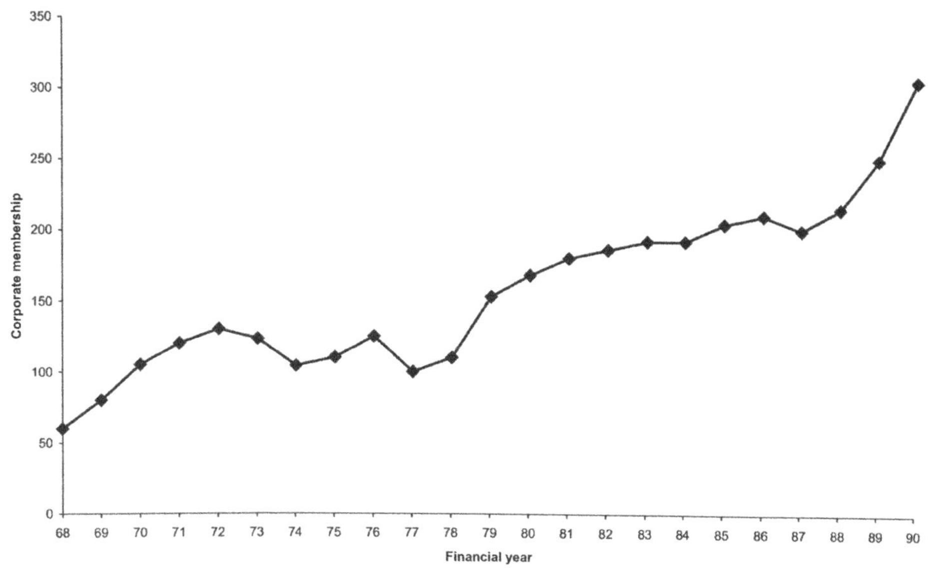

Source: SANF 1988, 1993

Figure 4.2 Corporate members 1968–1990.

conservation sponsorship, promoting the view that, by investing in the creation of a national park or reserve, companies would be making a bold gesture of faith in the future of South Africa.

The involvement of companies in SANF's nature conservation activities challenges recent attempts to explain TFCAs as a product of globalisation. For example, the creation of TFCAs is seen as result of regulated globalisation (Duffy 2001). The major flaws with these explanations are that they do not only offer a limited perspective of globalisation but also imply that the globalisation of nature is a recent phenomenon. From the perspective of the political economy, there is a rich history of the penetration of capitalism into nature conservation, including national parks (Harvey 1996; Langholz and Lassoie 2001; Emel 2002). In this sense, the trajectories of globalisation in TFCAs can more appropriately be understood by reflecting on the historical links between nature conservation and capital, and how the recently established TFCAs have added new dimensions to those links. While I endorse the view that TFCAs are connected to different trends of globalisation, I do not see TFCAs as a starting point for understanding those trends. The discussion in this chapter suggests that the connections between business and the current PPF has a long history that can be traced back to the establishment of SANF.

It could defensibly be argued that the late 1980s were crucial years for SANF. Indeed, the Chairperson of the Foundation, Gavin Relly,[15] did not only regard 1989 as the 21st anniversary of SANF, but also referred to it as 'a threshold of the future' (SANF 1989, 60), and the following year (i.e. 1990) SANF acknowledged that despite its successes over the last 22 years there was no guarantee of victory in the future. Against this backdrop, the Foundation's director at the time, Allan Heydorn, drew a strategic plan for 1989–1993, which had the following areas of focus:

(a) the preservation of genetic species and ecosystem diversity;
(b) ensuring the sustainable use of renewable natural resources;
(c) fighting wasteful consumption of resources and energy;
(d) combating pollution; and
(e) the promotion of responsible and caring attitudes by all people with the purpose of reversing the accelerating degradation in the planet.

In an article entitled 'Which way to the future?' SANF (1990, 34) referred to the existence of a master plan for conservation in these words: 'though some projects may focus on a specific small mammal, nondescript bird or little-known plant, they all tie together in a master plan for preserving our natural heritage as an *indivisible ecosystem*' (my emphasis). In the same year, Anglo-American and De Beers donated R2.6 million to SANF 'for urgent conservation projects' (SANF 1990, 45). That amount was said to be 'the largest unearmarked donation given by the corporate sector for conservation in southern Africa and one of the largest in the world' (SANF 1990, 45).

Implicit in the master plan was the Foundation's intention to join scattered game reserves into a single protected area, as was the case with the creation of Maputaland Coastal Forest Reserve. Together with WWF and the Oceanographic Research Institute, SANF also produced a conservation master plan for sustained development of Mozambique's Bazaruto archipelago in 1989. According to SANF, the plan was to safeguard the archipelago from the pressure of tourism and refugees fleeing the mainland (SANF 1993). The conservation projects along Mozambique's coastal areas were also supported by the International Wilderness Leadership School (WILD), and formed part of SANF's 1990/91 campaign to save ecosystems. WILD joined forces with Forum *Natureza em Perigo* and the successor of SANF, the PPF, to protect the *Reserva Especial* (Maputo Elephant Reserve), which falls within areas that were later to be used for establishing the Maputoland TFCA. In that way, conservation projects supported by SANF also became stepping stones towards the establishment of transfrontier parks. For example, in 1990, SANF carried out an ecological survey in Mozambique, the aim being 'to pave the way for the establishment of Africa's first transfrontier park, parallel and adjacent to the Kruger National Park'[16] (SANF 1991, 10).

The emergence of the idea of transfrontier parks within SANF was accompanied by two major interrelated developments, namely, the realignment of SANF's activities with the economic imperatives of the new black-controlled states and the restructuring of SANF. With almost all indications pointing to black majority rule in post-apartheid South Africa, SANF attempted to reorientate its nature conservation policies to include black communities in its core activities. This was clearly articulated in President Rupert's address to SANF members on 18 June 1991 in Stellenbosch, in which he commented that SANF would, henceforth, fund and encourage innovative projects, which placed emphasis on local communities. Such projects were to link nature conservation with economic development in order to make SANF's activities acceptable to the black people 'who previously dismissed [nature conservation] as a matter of low priority' (SANF 1991, 2). SANF aimed to sell the idea of transfrontier parks to black governments and their citizens on an economic rather than an ecological ticket. Apart from wanting to promote transfrontier parks as people-oriented, the economic advantages were touted to lure governments that were preoccupied with ways in which they could promote economic growth for the betterment of their citizens.[17] According to Dommisse and Esterhuyse (2005), Rupert envisioned that the African elephant would bring economic miracles in Africa in the same way that the Asian economic revival led to those countries being dubbed the 'Asian Tigers'.

The second development that accompanied the idea of transfrontier parks was the need to restructure SANF. According to the Chairperson of SANF, Gavin Relly (SANF 1991, 15), restructuring was important 'to enable the Foundation to handle rapidly increasing demands in the fields of environmental conservation, education and fund-raising … in view of political changes which [were] sweeping through the continent'. The appointment of John Hanks as the Chief Executive Officer of SANF in June 1990 was part of the restructuring of the Foundation. As a former director of Africa conservation programmes at the WWF in Switzerland, Hanks provided the much-needed experience of conservation in Africa, while, at the same time, reinforcing the Foundation's international links at a critical historical moment. He was, as the SANF report put it, 'the right man at the right time' (SANF 1990, 15). Hanks' experience of conservation in Africa was useful to SANF, which considered its conservation projects as a blueprint for conservation throughout Africa[18] (SANF 1987). His knowledge of elephants[19] was crucial to the future preoccupation of the Foundation with the establishment of elephant migratory routes under the theme of peace parks. He regarded TFCAs as one of the appropriate strategies for treating the underlying causes of threats to wildlife as it embraced biodiversity conservation as a key component of socio-economic development (*Africa Geographic* 2006). Against this backdrop, proposals for TFCAs and the restructuring of SANF were supportive to, and reflected, shifts in global environmental agendas.

It is not a coincidence that the restructuring of SANF occurred in parallel with those of global environmental institutions. For example, the proposal for the formation of GEF was made in the late 1980s, the same period that SANF was undergoing a major restructuring in terms of its administrative structure and strategies for nature conservation and funding. Some comments on the formation and aims of GEF are helpful for understanding the development of shared agendas between SANF and GEF. It is said that a banker, capitalist and conservationist, Michael Sweatman, promoted the idea of a green financial instrument in the 1980s. As the Director of WILD, his primary interest was to promote the private sector role in conservation worldwide, particularly in Africa (Young 2002). The idea of bringing capital and conservation together was appealing to governments and institutions that were under pressure from environmental movements and activists who had mounted a strong challenge against mainstream global economic development since the 1980s. 'By the mid 1980s', Young (2002, 35) wrote, 'a group of ten, mostly Washington-based environmental NGOs [such as the Sierra Club, Audubon, Wilderness Society, Natural Resources Defence Council, National Wildlife Federation, Environmental Defence Fund, National Parks and Conservation Association, and Friends of the Earth] gained greater access to the ears (and funds) of donor governments by suggesting that economic growth was not only compatible with environmental protection, but necessary to generate sufficient funds to pay for it'. That suggestion was in line with the 10-year review of the United Nations Environment Programme (UNEP), which recommended the creation of the World Commission on Environment and Development. In addition to furthering the idea of the compatibility of environmental and development interests, GEF was also a strategic response towards environmental activism on the one hand, and the need to generate funds for nature conservation on the other hand.

Following the publication of the Brundtland Report in 1987, the need for a green financial instrument became relevant to fund 'sustainable development'. At the same time, NGOs such as IUCN and WWF were lobbying for the creation of a few multilateral conventions that would bring together small and scattered agreements under one legal umbrella, in the form of international conventions. Against this backdrop, the first step towards establishing a green financial instrument was the creation of the International Conservation Financing Program (ICFP) in 1987 at the World Resources Institute in Washington, DC. The ICFP was not only led by an influential US economist, Robert Repetto, but its advisory panel included Sweatman and 'representatives of other NGOs, private investment banks and major development agencies (among them two World Bank directors)' (Young 2002, 51). The 1992 Rio Earth Summit also provided conditions under which the idea of the environmental facility could be carried forward. For example, the World Bank supported the creation of such a facility, viewing it as an avenue through which it could escape from controversies and criticisms surrounding its funding of certain projects. Some non-Western countries also aimed to use the Summit to press for the restructuring of the World Bank and for creating a green fund to compensate southern countries for the plundering of resources by northern countries. These developments highlighted that there was a need for a global green fund, though the purpose for that fund was debatable. GEF was formed to satisfy that need.

Of significance to the theme of this chapter is that some of the key figures in the formation of GEF, such as Sweatman, were also involved in institutions that were later to play a role in TFCA projects. As mentioned above, Sweatman was once a director of WILD, a foundation that was established in the United States in 1974 by South African Ian Player with the aim of protecting threatened wilderness areas and wildlife in sub-Saharan Africa.[20] Ian Player is a doyen of the wilderness movement that sought, and still seeks, to preserve and expand wilderness areas. He founded the Wilderness Trust in the United Kingdom and was a force behind the establishment of the World Wilderness Congress (Player 1997). In South Africa, he propagated the idea of wilderness[21] through the Natal Parks Board, which he served from 1952 to 1974. His Wilderness Leadership School is actively involved in the development of wilderness trails, including in uKhahlamba-Drakensberg Park, which forms part of the Maloti–Drakensberg TFCA. The symbol of his Wilderness Leadership School, the erythrina leaf, embodies the philosophy, 'Man to God, Man to Man, and Man to Earth', which conveys a biblical injunction of looking after the earth. Player's contribution towards preserving the wilderness was recognised by Prince Bernard of the Netherlands who honoured him as Knight in the Order of the Golden Ark (Player 1997). The point here is that there are points of convergence between actors in SANF, WILD and GEF.

The formation of GEF and the interests of the wilderness movement are relevant for understanding TFCAs in Africa, not least because GEF is one of the financiers of TFCA, as we shall see below. Whereas GEF has laid down the foundation on which TFCAs can be funded, the wilderness movement has ensured that preserving and protecting the wilderness for the enjoyment of tourists forms part of the TFCA agendas. The South African National Parks, as the main player in the development of TFCAs, has adopted the aspirations of WILD by introducing a Wild Card as part of its nature conservation funding formula, calling it 'an innovative subsidisation for conservation objectives'. I emphasise that developments leading to the formation of GEF and the objectives of WILD were reflected in, and perhaps even spurred on by the restructuring of SANF.

SANF's repositioning was short-lived, because it became WWF–South Africa in 1995 (Doolan 2006). As a South African chapter of WWF International, its activities were to be limited to South Africa in terms of the WWF's policy. Despite this policy, Rupert requested that WWF–South Africa be given special permission to operate outside South Africa, which the international WWF refused (Dommisse and Esterhuyse 2005). It has been reported that 'the Executive Committee of WWF South Africa concluded that a separate body was required to facilitate the process of TFCA establishment and funding' (PPF 1997). Subsequently, the PPF was created on 1 February 1997, with Rupert as its chairman. There is speculation that the first Executive Director of the PPF, John Hanks, was one of the main forces behind the establishment of the Foundation and the propagation of the idea of peace parks.[22] Nevertheless, Rupert emerged as the main figure behind the PPF. Accordingly, the establishment of TFCAs has been considered one of Rupert's grandest and most ambitious schemes of his life (Dommisse and Esterhuyse 2005).

Of importance to our discussion in this chapter is the manner in which the PPF drew on the experiences of SANF. Arguably, the skills, resources and connections that existed during the life of SANF were transferred to the newly established PPF. For example, some of the key figures in the defunct SANF, such as Rupert, Hanks, Relly and Stroebel, became directors of the PPF in 1997. In the same vein, the PPF inherited some of the corporate founder members such as De Beers Consolidated Mines Limited. Other corporate founder members in 1997 were Billiton PLC (United Kingdom), Kransnapolsky Hotels and Restaurants/ The Lord Charles Hotel (the Netherlands) and Vendôme Luxury Group (Switzerland). The six major donors (Chase Manhattan Foundation, De Beers, Prince Bernard, PSP Publishing, Rupert Nature Foundation and Sanlam) contributed more than R100,000 each to the Foundation in 1997 (PPF 1997). Unlike its predecessor (i.e. SANF), which limited its membership to companies, the PPF drew its individual founder members from the Peace Parks Club that was also established in 1997. As Table 4.2 shows, South Africa had the highest number (36) of disclosed individual founder members in the first year of the Foundation. The following year (i.e. 1998) South Africa took the lead with 17 corporate founder members[23] of the Peace Parks Club compared to the Netherlands, which had one member,[24] Switzerland[25] (2), United Kingdom[26] (1) and the United States joining in with one member[27] only. In this way the PPF inherited SANF's impressive capital base. More importantly, conservation projects that had been initiated by SANF were useful in the actual establishment of TFCAs. In the section that follows I analyse some of SANF's activities that turned into important sites for TFCAs.

SANF's projects as TFCA sites

The fact that the PPF earmarked the establishment of seven TFCAs (Kalahari, Ais-Ais/ Richtersveld, Kruger/Banhine–Zinave/Gonarezhou, Gariep, Maputaland, Drakenberg/Maloti and Dongola/Limpopo Valley) in its first year of operation (i.e. 1997), suggests that the sites for TFCAs were known before the creation of the PPF. In its first year of operation, the PPF estimated that the funds needed to get the seven TFCAs off the ground would be R45,000,000 for Dongola/Limpopo Valley (the current Limpopo/Shashe), R80,130,000 for Drakensberg/Maloti, R10,200,000 for the Gariep, R39,600,000 for Kgalagadi, R20,700,000 for Maputaland, R5,300,000 for Kruger/Banhine-Zinave/Gonarezhou, and R7,900,000 for Ais-Ais/Richtersveld (PPF 1997). The PPF thus adopted a multi-site approach towards the establishment of TFCAs in southern Africa. This was possible because SANF was involved in most of the areas in which TFCAs were to be established.

In the 1970s, the Foundation supported the concept of a single conservation area in the northern part of the then South West Africa, the intention being to link the Skeleton Coast and Etosha National Parks with Kaokoland and Damaraland 'to protect the unique wildlife communities of the region' (SANF 1986, 31) and to 'promote Kaokoland as one of the

Table 4.2 Individual membership of the Peace Parks Club per country, 1997.

Country	Number of members
Bahamas	1
Belgium	1
Bermuda	1
China	5
France	4
Germany	4
Greece	1
Italy	5
Mozambique	1
Norway	1
Pakistan	1
Spain	2
South Africa	36
Sweden	1
Switzerland	30
The Netherlands	7
United Kingdom	20
United States	14
Zimbabwe	1

Source: Compiled from PPF 1997

world's top wilderness areas' (SANF 1986, 25). By 1986, SANF had committed an amount of R172,609 towards the achievement of these goals. That area has become a core of the PPF-proposed Iona–Skeleton Coast TFCA between Angola and Namibia.

SANF's first nature conservation project was in Swaziland, where it had donated R47,224 towards the enlargement of Mlilwane Game Reserve[28] in 1968 (SANF 1993),[29] the year Swaziland became independent from British rule. By 1982 SANF had committed R125,000 to nature conservation projects in Swaziland (SANF 1984). As I have intimated above, SANF was involved in the translocation of wildlife to Swaziland. It also financed the creation of nature reserves in the Swazi-dominated South African 'homeland' of KaNgwane. It committed R100,000 towards the creation of the present-day Songimvelo nature reserve[30] in 1984, which it considered then as the third largest nature reserve in South Africa after the Kruger National Park and the Kalahari Gemsbok National Park (SANF 1986). In 1986, Chris Mulder Associates produced a Master Plan and Management Guidelines for the Songimvelo Natural Resource Area (SNRA). The reason for establishing SNRA had very little, or nothing, to do with an insufficient amount of land designated as a protected area, because it was estimated that KaNgwane had about 24% of its total surface area under conservation in the mid-1980s. The SNRA was established primarily to protect the watershed that provides a high proportion of streamflow to major tributaries of the Nkomati River, which crosses the boundaries of South Africa and Swaziland. The consultants conceded that 'the amount of water generated in the SNRA serves as one of the most important reasons why the area should be conserved' (Chris Mulder Associates 1986, 83). That view was also endorsed by the Department of the Chief Minister of KaNgwane, which maintained that, 'Songimvelo is … vital for water catchment for other areas in KaNgwane, and its conservation … is thus a priority' (KaNgwane 1993).

The involvement of SANF in Swaziland and the former KaNgwane laid the foundation for the establishment of the PPF-driven Songimvelo–Malolotja TFCA. According to the PPF (2002, 5) 'the area has tremendous potential to be developed as an international peace park with unique and rare qualities that will make it a popular tourist destination'. The PPF defended its involvement in the Songimvelo–Malolotja TFCA on the grounds that the area is relatively small and therefore does not attract funding from other major donors, and because it will take no longer than 2 years to establish because of the excellent political relations between South Africa and Swaziland. Its other reasons were that the area has unique attributes and characteristics for it to obtain an international peace park status, and it offers job creation opportunities and cultural ecotourism to surrounding rural communities that are completely dependent on agriculture (PPF 2002). The establishment of the Songimvelo–Malolotja TFCA was also facilitated by the GEF project in Swaziland, the Biodiversity Conservation and Participatory Development (BCPD), which aims to conserve biodiversity and tourism corridors of which the TFCA would form part. The ultimate goal for developing Songimvelo–Malolotja TFCA was to incorporate it into the Lubombo TFCA.[31]

On the Mozambican side, SANF was involved in the establishment of the Maputo Elephant Reserve. It also supported conservation projects that were run by the Natal Parks Board on the South African side. SANF had conservation projects along the borders of Mozambique, South Africa and Swaziland. The PPF used those projects to launch the Lubombo TFCA. It should be noted that the Lubombo TFCA is also linked to the Lubombo Spatial Development Initiative (LSDI) that was officially launched in May 1998. There are speculations that the Lubombo TFCA resulted from the Spatial Development Initiative (SDI).[32] The heads of the three participating states at the time described the objectives of the LSDI in these words:

> We are putting in place a stable environment to regulate and facilitate investment. It will define and ensure forms of secure property ownership, protect and conserve the environment, ensure that local communities benefit from development in their areas and enable the private sector to maximise the many opportunities that exist (Mandela 1998).
>
> The initiative puts aside the barriers to trans-national co-operation in order to end the paradox of poverty amidst plenty (Chissano 1998).
>
> Our governments understand fully that the protection of the Lubombo's great natural beauty is a key to sustained investment and growth (King Mswati III 1998).

Subsequently, the three countries entered into the General Protocol on the Lubombo Spatial Development Initiative on 5 July 1999, paving the way for the linkages between the LSDI and TFCAs. Clearly, the LSDI was meant to create an investment-friendly environment to facilitate the development of infrastructure across the borders of the three countries. The aims of the LSDI are to: promote public sector investment in improvements to the region's infrastructure, mainly by improving transport corridors;[33] establish a small business support programme with relevant agencies that will allow local residents to take up opportunities for new commercial activity along the transport corridor; create an attractive and stable climate for investors in which to operate; develop and support a transnational malaria control programme; prepare and package opportunities for private-sector investment in tourism and agriculture; broaden ownership patterns in the regional economy; negotiate with institutions to secure affordable loans for local communities to take joint ventures with outside investors; and build an international competitive regional economy (Lubombo Spatial Development Initiative 1998).

As in other development projects, conservationists were concerned about the impact of development projects on the environment. Nevertheless, proponents of TFCAs considered the development of infrastructure, such as roads, as important for cross-border tourism. Accordingly, the following projects were attached to the LSDI: the Greater St Lucia Wetland Park; Lake Sibayi; Pouto do Ouro–Kosi Bay Transfrontier Tourism Project; Lavumisa–Jozini Transfrontier Tourism Project; Ndumu–Tembe–Futi–Usutu Transfrontier Concept; and Hlane–Mlawula Complex Transfrontier Tourism Project (Lubombo Spatial Development Initiative 1998). Arguably, these projects predate the launching of the LSDI; their focus was adapted to the LSDI in order to promote the establishment of TFCAs, with cross-border tourism as an important stimulant. Indeed, the General Transfrontier Conservation and Resource Area Protocol (2000, 3) that was signed by Mozambique, South Africa and Swaziland on 22 June 2000 has its first objective as the 'economic development through appropriate optimisation of opportunities presented by the Parties' natural assets'. A full discussion on TFCAs and tourism is given in Chapter 7. For the moment our focus is on the development of TFCAs in areas that had the attention of SANF and, later, the PPF.

Unlike other TFCA projects in southern Africa, the Lubombo TFCA developed as a complex of nature conservation initiatives that were linked together across the borders of the three countries. The Lubombo TFCA is a constellation of five TFCAs, namely, Ponto do Ouro–Kosi Bay (Mozambique and South Africa), Usuthu–Tembe–Futi (Mozambique, South Africa and Swaziland), Nsubane–Pongola (South Africa and Swaziland), Lubombo Conservancy–Goba (Mozambique and Swaziland) and Songimvelo–Malolotja (South Africa and Swaziland) (Figure 4.3). The protocols for the first three TFCAs[34] in Lubombo were signed on 22 June 2000. The signing of protocols for the various TFCAs across the Lubombo Mountains on the same day is further evidence that the raw material for TFCAs had long been in place. That is to say that the TFCAs on the Mozambique–South Africa–Swaziland border were established on conservation projects that had been supported by SANF, hence those TFCAs have developed simultaneously on different sites. It should be noted that the broad aims of the TFCAs in the Lubombo are economic development, transfrontier ecological planning and resource management, and 'the supporting and strengthening of existing and proposed initiatives underway in respect of these objectives in the Region and the creation of an enabling framework to assist such initiatives and to create synergies between and among them' (General Transfrontier Conservation and Resource Area Protocol 2000, 4). The Protocol therefore recognises most of the projects that SANF and other environmental NGOs had undertaken in the area and supports the use of those projects in launching the TFCA initiatives.

The Maloti–Drakensberg TFCA on the Lesotho–South Africa border developed along similar lines. SANF financed the establishment, fencing[35] and supply of equipment and research facilities for Lesotho's first national park, Sehlabathebe, created in 1970 (SANF 1984). It should be noted that South Africa had influence over the development of Sehlabathebe long before the emergence of TFCAs. According to Nyankale (1979), the first

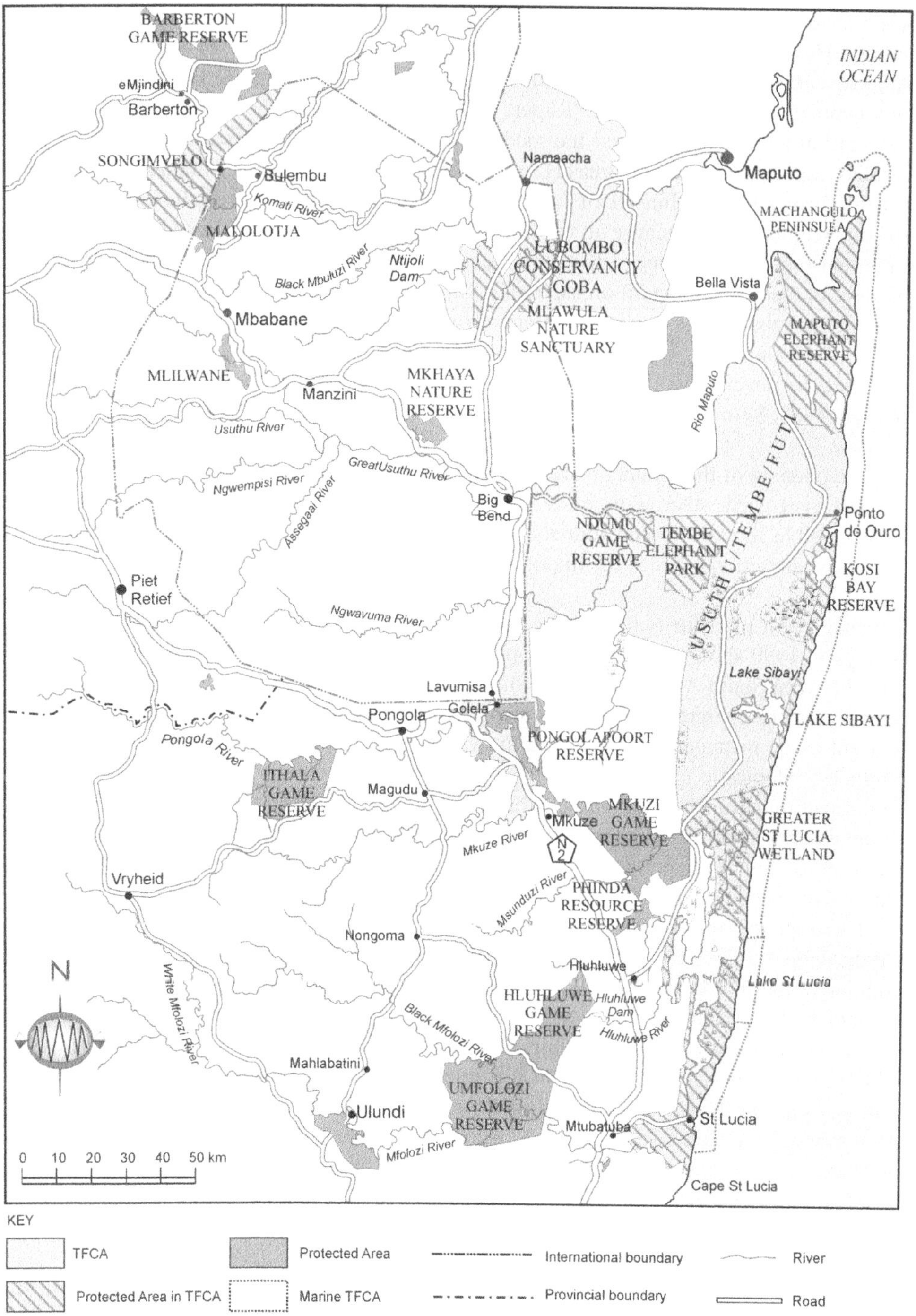

Source: Compiled from various sources

Figure 4.3 The Lubombo TFCAs.

proposal for Lesotho's national park was the Oxbow area in the north-western part of that country. However, following the on-site visit by the Deputy Director of the South African National Parks Board on 8 May 1969, it was decided that Sehlabathebe should be Lesotho's first national park. At the time, Rupert had already created the Lesotho Development Corporation and had established the medical shuttle that provided free medical service to Lesotho, especially in remote areas (*Cape Times* 2006a; *Sunday Times* 2006a). It is alleged that the former Prime Minister of Lesotho, Jonathan Leabua, endorsed the creation of the park at Sehlabathebe, because it was an ideal hideout place.[36] The Lesotho Government supported the establishment of the park, because it fitted into the government's policy for the development of mountain areas, which aimed at making those areas accessible to the lowland markets through the development of roads. It commissioned the Transport Research Centre of the University of Stellenbosch to determine the road network. The ultimate goal for infrastructure development in mountain areas was clearly captured by former Prime Minister Leabua (1968, 3) in these words:

> The opening of the mountain areas will constitute the beginning of tourism. People will visit each other with ease. Handicrafts made by mountain people will be accessible to visitors and tourists. The countryside will itself constitute a major tourist attraction – hence the importance of trees, and fish in the rivers and dams.

A management plan for Sehlabathebe that was developed by the Directorate of Overseas Surveys in 1990 shows that there were plans to officially recognise the migratory route of game between South Africa and Lesotho. Though one of the objectives of the management plan was 'to prevent uncontrolled movement of game across the national park boundaries into adjacent range management areas, and the violation of national park boundaries by illegal entrance of domestic animals from the range management areas,' it was emphasised that 'movement of game between the Sehlabathebe National Park and the adjacent Drakensberg Mountain Park should not be hindered' (Lesotho 1990, 11). In other words, the management plan suggested that the boundary fence between the then Drakensberg Park (South Africa) and Sehlabathebe National Park (Lesotho) would be removed.

The impetus for a TFCA in the Drakensberg came from the Natal Parks Board, which had the support of SANF and the South African government, especially the Natal Provincial Administration. For example, R.W. Hicks compiled the final conceptual development plan for Sehlabathebe National Park, and the entire planning project was funded by the South African Department of Foreign Affairs (Hicks 1991). The plan was to manage and market Sehlabathebe National Park as part of the South African Drakensberg tourist product. That is to say that the Natal Parks Board incorporated Lesotho's national park into the Natal Drakensberg Mountain Park's tourism development policy. Apart from its tourism potential, the Drakensberg is a water catchment area for South Africa's three provinces (Gauteng, Free State and KwaZulu-Natal) and Lesotho. Indeed, the water catchment area of the Drakensberg has been central to South Africa's diplomatic relations with Lesotho, hence the two countries signed the Lesotho Highlands Water Treaty in 1988.

The development of the Drakensberg TFCA can be traced back to the Natal's Town and Regional Planning Report of 1990,[37] in which the Provincial Administration issued a statement of policy in the form of the Drakensberg Approaches Policy (DAP). The DAP provided 'a comprehensive planning and management strategy throughout the region, to control further development, provide recreation and promote tourism in the national interest'

(Town and Regional Planning Commission 2001, 9).[38] Viewed from the perspective of political developments in South Africa in 1990, DAP was part of the strategy to pre-empt a regional dispensation in which part of the former 'white' area of Natal would remain under the control of the white minority. However, contestations over the demarcation of regional boundaries for the whole country (see Muthien and Khosa 1998; Ramutsindela and Simon 1999) meant that DAP was politically inappropriate. Subsequently, Willem Van Riet Landscape Architects were commissioned to determine land use zoning in the Drakensberg in 1993, and were again appointed in 1994 to formulate a Development Control Scheme for the uKhahlamba district of the former 'homeland' of KwaZulu (Town and Regional Planning Commission 2000). A breakthrough in the control of the Drakensberg came in 1996/97 when the Town and Regional Planning Commission initiated the Drakensberg Special Case Area (Town and Regional Planning Commission 2001).[39] 'A Special Case Area (SCA) is an area which, because of its unique character, requires additional protection beyond the development application procedures which apply in any event' (Town and Regional Planning Commission 2001, 1). The need for designating the Drakensberg as a Special Case was defended on the basis that it was under threat from four main sources: tourism development; commercial farming; commercial afforestation; and rural communities. It was concluded that 'the existing institutional structure in the Drakensberg, suffering from the same ills as the rest of the province, is not competent to deal with the problems in the Drakensberg because of the general development issues, in particular, the importance of the Drakensberg as [a] primary water catchment area and international tourist destination' (Town and Regional Planning Commission 2000, 31). It should be noted that efforts to manage the Drakensberg as a special case area were limited to the South African side of the mountain. The Special Case Area Plan for the Drakensberg proposed seven land use zones (Table 4.3) on the South African side of the Drakensberg, using the uKhahlamba–Drakensberg Park as the core area.

Table 4.3 Proposed land use zones for the Drakensberg as a Special Case Area.

Zones	Estimated area (ha)	% of SCA
uKhahlamba–Drakensberg Park	228,250	40.0
Buffer	71,600	12.5
Commercial agriculture	182,650	32.0
Resource conservation	47,240	8.3
Settlement in buffer	3,540	0.6
Settlement in the commercial/ agricultural zone	19,180	3.4
DAP nodes and termini	18,540	3.2

Source: Town and Regional Planning Commission 2001

When the concept of the Drakensberg TFCA was made in 1997, the initial proposal was to link the uKhahlamba–Drakensberg Park (South Africa) with Sehlabathebe National Park (Lesotho) (Town and Regional Planning Commission 2000; Sandwith 2003). It should be emphasised that the Town and Regional Planning Commission wanted to incorporate Sehlabathebe National Park as part of the Transfrontier World Heritage Site together with uKhahlamba–Drakensberg Park. According to Sandwith (2003), South Africa and Lesotho

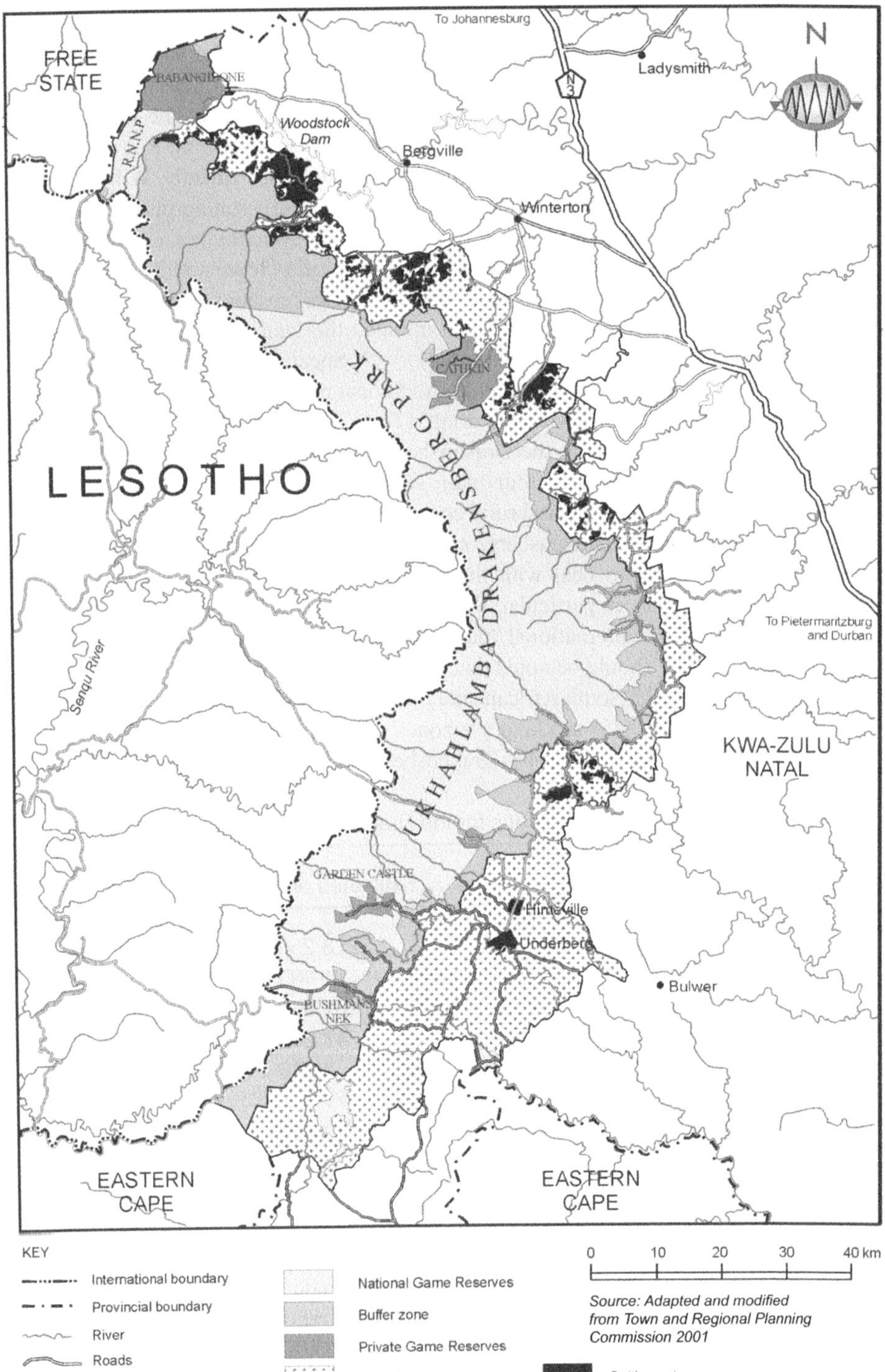

Figure 4.4 The South African side of the Maloti–Drakensberg TFCA.

agreed to establish a co-operative transboundary conservation and development area after a workshop involving a bilateral delegation and international donors, which was held in September 1997.[40] Though the processes of establishing the SCA and the World Heritage Site over the Drakensberg occurred in tandem, the end result was that the UNESCO's World Heritage Committee inscribed uKhahlamba–Drakensberg on the World Heritage List in 2000 in terms of natural and cultural criteria (UNESCO 2000). Clearly, the SCA initiative played a significant role in the zoning of the heritage site, leading to the development of the boundary around the SCA on the South African side only (Figure 4.4). In this way, the boundary of the Maloti–Drakensberg TFCA first developed on one side of the international border, where it was influenced by land use in the present-day province of KwaZulu-Natal. The surprisingly smooth boundary of the TFCA on the Lesotho side was drawn arbitrarily, and in haste, as Sehlabathebe National Park was too small to match either the size of uKhahlamba or the area that had been envisaged for the TFCA.

The development of TFCAs in the Drakensberg and Lubombo mountains reveals that existing conservation activities contributed to the development of those TFCAs. Nevertheless, the process of consolidating disparate conservation projects into TFCAs was a complex undertaking, not least because there was no model for TFCAs in the region or the African continent. More importantly, conservation areas that were to be consolidated into TFCAs were not contiguous but most were, and still are, separated by human settlements. The result was the emergence of TFCAs that were known as transfrontier conservation and development areas. Their designation as conservation and development areas signifies the attempt to pursue development and conservation objectives together in the same areas. The first model for TFCAs came in the form of a transfrontier park.

The rise of transfrontier parks

Existing local conservation initiatives largely shaped the geography of transfrontier parks in southern Africa. This is so because transfrontier parks are founded on protected areas, especially national parks that are simply joined together across national boundaries in line with the Peace Park model discussed in Chapter 3. Different reasons have been advanced for the need for transfrontier parks. Firstly, transfrontier parks are promoted on ecological grounds: national parks should form a network of protected areas that would be supportive to the re-establishment of ecosystems, which are crucial to the survival of humans and all species. That is to say that transfrontier parks aim to perfect the ecological roles that national parks were meant to serve, but that they failed to perform, mainly because they were initially established to serve different interests, some of which had nothing to do with the protection of biodiversity. The ecological aim of transfrontier parks is therefore to re-establish ecoregions, using national parks as raw material in that process. In this way the boundaries of national parks are being used as that of an ecoregion. The only difference is that the international boundary that separates, say, two adjacent national parks is removed so as to turn two national parks into a single ecological unit. This is problematic because the outer boundaries of national parks, which were not founded on bioregional planning, remain the same. Moreover, the establishment of transfrontier parks, especially in Africa, was not preceded by the definition of ecoregions. It should be emphasised that the popular view is that the existence of contiguous national parks globally suggests the possibility of establishing hundreds of transfrontier parks across the globe (see Chapter 3). That estimate is based on the number of contiguous national parks rather than on the ecoregional map of the world. The implication of national parks-based-ecoregions is that they perpetuate ill-defined ecoregions while, at the same time, encouraging the establishment of more protected areas along international borders so as to promote the establishment of transfrontier parks. In this way transfrontier parks become ecoregions that are largely defined at the international scale, and therefore contrast sharply with bioregionalism as a social movement that aims at the local control of natural resources (see Chapter 2). The contrast is not only in terms of scale (i.e. local versus international), but is mainly in what constitutes a bioregion. By relying on national parks as pillars for ecoregions, transfrontier parks promote large-scale bioregions that are devoid of human settlements, and the boundaries of such ecoregions are defined mainly on the basis of species behaviour. These kinds of bioregions lead us to the second reason for the establishment of transfrontier parks, the promotion of ecotourism.

The long history of tourism in national parks is key to the establishment of transfrontier parks. As I have intimated above, SANF aimed at establishing Africa's first transfrontier parks in and around South Africa's famous Kruger National Park, because of the high number of tourists[41] visiting that park every year. The desire to use transfrontier parks to recreate the wilderness for ecotourism accounts for the hostile attitude that the promoters of transfrontier parks have towards human settlements in areas designated for those parks. The plan for transfrontier parks is to allow tourists to visit expanded national parks without the restrictions that were initially imposed by border controls. Indeed, the potential economic benefits from the flow of tourists have enticed the governments of participating countries while, at the same time, rendered local populations less resistant to the establishment of transfrontier parks (see Chapter 6). It should be noted that the establishment of transfrontier parks in the southern part of Africa did not occur in a vacuum, but depended heavily on isolated conservation initiatives, a factor that accounts for their differentiation, as I will show in the section that follows.

The Kgalagadi: Africa's first transfrontier park by default

Indications are that although the Kgalagadi Transfrontier Park (KTP) was the first to be established in the region and the continent as whole, it was not at the top of the agenda of advocates of transfrontier park projects. Two reasons can be advanced to support this view. Firstly, the geographical spread of SANF's more than 200 conservation projects did not include any part of the area that falls under the KTP. SANF's interest in Botswana was mainly in the central and northern part of that country. For example the Foundation was involved in the translocation of white rhino from Natal to Chobe National Park and also financed a 4-year study of lechwe antelope in the Okavango Delta (SANF 1984). Secondly, as we have noted above, SANF proposed to develop Africa's first transfrontier park in and around the Kruger National Park. Official statements suggest that the proposal to establish the KTP resulted from Botswana's commissioned fact-finding study on tourism that was conducted in 1989. Subsequently, the Department of Wildlife and National Parks in Botswana consulted with the South African National Parks in 1991, leading to the establishment of the Transfrontier Management Committee in 1992 (De Villiers 1999). The Committee was responsible for drafting the proposal for the establishment of the transfrontier park. This begs the question of why the Kgalagadi became the first transfrontier park in the region. The answer can most probably be found in the history of the two parks that were joined together to form the KTP, the nature of state borders separating the two parks, and the working relationships between conservation agencies in South Africa and Botswana.

The history of protected areas in the Kgalagadi reveals that governments were more interested in territorial control than in nature conservation. Before the race to control the territory of the Kalahari began, both the colonial government in the Cape and the British government were reluctant to commit themselves to the Kalahari. British interest in the Kalahari was strongly influenced by the desire to ward off German imperialism in the western part of southern Africa. Even after the occupation of Bechuanaland by Britain, the colonial government in the Cape remained non-committal to Bechuanaland up to the time of the replacement of John Mckenzie by Cecil Rhodes as Deputy Commissioner (Colonial Office 1884). Despite their political interests in Bechuanaland and their active participation in the establishment of national parks, the British were not interested in establishing a national park in the Kalahari. The first national park to be created there, the Kalahari Gemsbok National Park, was actually established by the Union of South Africa (on the South African side) in 1931. Its counterpart in present-day Botswana was established 7 years later (i.e. 1938), with strong influence from the South African side. There had been co-operation between the South African and Botswana conservation agencies since the creation of the two national parks. De Villiers (1999) states that co-operation in the form of a 'gentleman's agreement' had been in place since 1948. Against this backdrop, the formation of the Transfrontier Management Committee in 1992 was the formalisation of a structure that had been in existence for almost half a century.

It follows that the groundwork for the establishment of the KTP was laid by the Department of Wildlife and National Parks in Botswana and the South African National Parks (SANParks). However, the formalisation of the agreement between these two agencies required the approval of the governments, particularly in South Africa where SANParks is not a government department, but simply a statutory body. South Africa signed the agreement on 3 March 1999, with Botswana following a month later on 7 April 1999. The governments of the two countries simply endorsed the Management Plan that had been agreed to between Botswana Department of Wildlife and National Parks and SANParks. The Kgalagadi Transfrontier Agreement (1999, 6) acknowledges that 'the management objectives shall, in the first instance, be implemented through the [Botswana] Wildlife Department and [SANParks] concluding a Record of Understanding for a detailed Management Plan'. The pivotal roles that SANParks and the Wildlife Department played in the creation of the Kgalagadi accounts for the absence of an international co-ordinator[42] during the creation of that park. There was no need to invoke the relevant SADC or Global protocols to persuade the two countries to sign the agreement, because the agreement was a foregone conclusion.

Against this backdrop, the PPF played an indirect role during the formative stages of the Kgalagadi. Nevertheless, SANParks and the Director of the Department of Wildlife and National Parks in Botswana, Mr Sedia Modise, represented the interests of the PPF in the establishment of the Kgalagadi. This claim is made on two main grounds, the first being that SANParks, as a conservation agency for the South African government, is responsible for all national parks in the country. Decisions about national parks have to be endorsed by SANParks. Furthermore, the first group of almost all international co-ordinators for TFCAs came from SANParks, suggesting that high-ranking personnel in the conservation agency supported the idea of TFCAs. The second reason is that Mr Sedia Modise is, according to the PPF Newsletter, one of the principal architects of the KTP (PPF 1999). He is also a founding member of the KTP Foundation (KTPF) that was established in 1999 to direct the activities of the KTP. The KTPF was biased towards South Africa from the onset. For example, it was created as a Section 21 Company under the South African Companies Act of 1973; the South

African Reserve Bank was to authorise the disbursement of funds; and SANParks provided the secretarial services to the Foundation (Kgalagadi Transfrontier Agreement 1999). In 2002 Mr Modise was the co-author of the general proposal for TFCAs in southern Africa. By 2003 he was in charge of the PPF's first external satellite office in Botswana.[43]

The path towards the establishment of the KTP was also smoothed by the absence of a border fence between South Africa and Botswana. Unlike most state borders in Africa, small blocks of cement mark the border between South Africa and Botswana along the Nossob River, which is generally dry. That is to say that there is no fence separating the two national parks along the international boundary. More crucially the Kgalagadi was the only site between South Africa and its neighbours where national parks existed back to back, without the interruption by any other form of land use. A combination of these factors provided the fertile ground for experimenting with transfrontier parks in the Kgalagadi. Arguably, the two national parks in the Kalahari existed as a de facto transfrontier park in terms of their geography and management structures.

The brief history of the KTP sketched above supports my claim that transfrontier parks define ecoregions on the basis of the borders of national parks. Article 1 of the Kgalagadi Transfrontier Agreement (1999, 2) between Botswana and South Africa on the KTP commits the two countries to recognise the Kalahari as 'an undivided ecosystem'.

Clearly, the use of the borders of existing national parks to define the extent of the ecosystem has compromised the ecological integrity of the Kalahari, which goes as far as Namibia. However, Namibia was excluded from the KTP, because it was not party to the pre-transfrontier park co-operative arrangements and the political relationship between Namibia and South Africa in the early 1990s was still to be redefined: the consequential effects of South Africa's colonial practices in Namibia (then South West Africa) since the end of World War I had still to be addressed. More crucially, Namibia was excluded from participating in the KTP for convenience; proponents of transfrontier parks in the region were desperate to prove the possibilities of establishing transfrontier parks in Africa, more especially because TFCA projects were proving to be difficult to establish. In this sense, the establishment of the Kgalagadi provided much-needed relief, as the following quotations show:

> It is a great honour to be here today on the border of Botswana and South Africa as we make history with the opening of the first transfrontier park in Southern Africa and one of the first transfrontier parks established in the world by legal statute … I am also pleased to add that the opening of the Kgalagadi is not the end of the road but a *milestone in our plans* for other transfrontier conservation areas in southern Africa … (Moosa 2000).

> A very important breakthrough for the growth and development of peace parks in the rest of Africa (Peace Parks Foundation 1999, 1).

Implicit in the quotations above is the view that the establishment of the KTP served as a guideline for similar projects in the continent. The Chief Executive Officer of the Kalahari Conservation Society (KCS), Wazha Tema, emphasised that view at the launch of the KTP on 12 May 2000 when saying that the establishment of KTP is to the KCS 'a historic event in that it is the first Transboundary Park to be officially established and launched through a formal treaty. The two governments have pioneered an environmental management strategy, which should be envied by most countries, particularly African countries' (Tema 2000, 1).

Indeed, the experience gained in the creation of the KTP has been used in the creation of other peace parks in the region. The KTP has provided proponents of TFCAs with a useful tabloid on which to draw agreements for the establishment of TFCAs. More importantly, it has reinforced the place of national parks as a cog in the development of TFCAs. Hanks (2003) acknowledged that the KTP provided a model for establishing TFCAs. In practice, the model suggests a shift from the focus on TFCAs to transfrontier parks, hence there has been an upsurge in the creation of national parks along state borders so as to facilitate the development of such parks, as has been the case with the Great Limpopo, Ais-Ais/ Richtersveld and the proposed Limpopo/Shashe Transfrontier Parks.

Creating the world's biggest animal kingdom: the Great Limpopo

As I have intimated above, protagonists of transfrontier conservation projects in Africa, particularly SANF, envisioned that Africa's first transfrontier park would be established on the Mozambique–South Africa–Zimbabwe border. However, Mozambique and Zimbabwe did not have national parks that could be joined directly with South Africa's Kruger National Park. That is to say that national parks could not be used as the core of the envisaged transfrontier parks, because there were no contiguous national parks on the proposed site. The situation was complicated by the absence of a national park on the Mozambican side that was adjacent to the Kruger National Park while Gonarezhou National Park in Zimbabwe was separated from Kruger National Park by human settlements.[44] Consequently, there was some confusion whether the conservation project should be a transfrontier park or a TFCA or both. In practice, conditions at the proposed site (i.e. on the borders of the three countries) favoured the establishment of a TFCA rather than a transfrontier park. However, establishing a TFCA would have defeated the intention of the PPF to create peace parks – as TFCAs are not parks. Unsurprisingly, the PPF pursued the idea of creating a transfrontier park between the three countries. The first step towards that goal was to create a new national park, Limpopo, on the Mozambican side in 2001 so as to join the area of Mozambique with the Kruger National Park. Joining Gonarezhou with the parks in Mozambique and South Africa remained problematic, because human settlements form a strip between Gonarezhou and national parks in the other two countries. The need for that strip was formalised in the Memorandum of Understanding (1999), the Trilateral Agreement (2000) and finally the Treaty that was signed in 2002. Both legal documents accentuated the creation of a park above that of a TFCA, thereby reinforcing the Waterton/ Glacier Peace Park model. The PPF (2000, 1) issued the following statement to celebrate the signing of the Trilateral Agreement:

> Now with the signing of the International Transfrontier Conservation between the three countries on Friday 10 November 2000 to develop the Gaza/Kruger/ Gonarezhou Transfrontier Park, the dream has finally been realised. This agreement, when fully implemented, will consolidate nearly 100,000 square kilometres of land – an area bigger than Portugal – into one of the world's largest wildlife parks, a model for conservation in the 21st century.

It should be noted that the PPF referred to the establishment of the KTP as 'a breakthrough', but considers the Gaza–Kruger–Gonarezhou Transfrontier Park as a model for cross-border conservation projects. This distinction suggests that the PPF has been preoccupied with the establishment of the GLTP, which it wants to showcase as its biggest achievement. The GLTP became a site on which the re-establishment of ecological systems across state borders could physically be demonstrated to a wider audience. It is precisely where the fence was, for the first time, literally cut in the establishment of transfrontier parks in southern Africa.

A closer reading of the Treaty of the GLTP reveals that the establishment of that park set a new tone in the development of transfrontier parks in the region. Unlike in the Kgalagadi where no relevant protocols are referred to, the Treaty of the Great Limpopo invokes both regional and global conventions and protocols so as promote the international credibility of transfrontier parks, to bind states to, and to insert these parks into, high-level governance structures. The specific conventions and treaties that were used as a background to the Treaty of the GLTP are the Convention on International Trade in Endangered Species of Wild Flora and Fauna (CITES) (1973), Convention on Biological Diversity (1992), Treaty of the Southern African Development Community (SADC) (1992), and the United Nations Convention to Combat Desertification (1994). It should be noted that the Treaty did not refer to any convention or protocol that deals specifically with humans – the evidence that the creation of wilderness, rather than meeting socio-political needs and aspirations, was paramount. As I will be showing in Chapter 5, these legislative instruments were useful in taking the task of establishing transfrontier parks beyond the territories of the state. The GLTP reconfirmed the importance of national parks in transfrontier conservation and provided a solid foundation on which other transfrontier parks could be established in the region and the African continent. This is evident in the Ais-Ais/Richtersveld Transfrontier Park (ARTP) on the Namibia–South Africa border that was established by a treaty in 2003, a year after the signing of the Treaty of the GLTP.

The Ais-Ais/Richtersveld Transfrontier Park

Apart from the Foreword of the Treaty of the ARTP, which is exactly the same as that of the GLTP, the ARTP was founded on regional and international conventions. These include the Waterfowl Habitat (1971), the Convention Concerning the Protection of the World Cultural and Natural Heritage (1972), the Convention on International Trade in Endangered Species of Wild Flora and Fauna (1973), the Convention on Biological Diversity (1992), the Treaty of the Southern African Development Community (1992) and the United Nations Convention to Combat Desertification (1994). As in the case of the KTP and the GLTP, protected areas – Ais-Ais Hotsprings Game Park (Namibia) and the Richtersveld National Park (South Africa) – were used to launch the ARTP. However, unlike the first two transfrontier parks in the region, the ARTP did not only make reference to the convention on the protection of cultural rights, but also moved away from the general concept of parties to that of stakeholders in order to assert the rights of local communities, particularly on the South African side where the national park land belongs to the Richtersveld Community. Article 1(j) makes this point clear: '"stakeholders" means individuals or groups of individuals, communities or representative institutions with an enforceable *right in land*, which included in the Transfrontier Park, and shall include the Richtersveld Community' (Treaty of Ais-Ais/Richtersveld Transfrontier Park 2003, 6). Whether those rights would make the Richtersveld Community any better off than communities in other transfrontier parks is debatable. Nonetheless, the fact is that the ARTP has introduced a new dimension of the community into the establishment of transfrontier parks.

Limpopo/Shashe

The fourth transfrontier park on the line in southern Africa, the Limpopo/Shashe on the Botswana–South Africa–Zimbabwe border, is set to consolidate the model that was set by the Kgalagadi and perfected by the GLTP and ARTP. Notably, the Limpopo/Shashe[45] was one of the first TFCAs to be earmarked by the PPF. As in the case of the GLTP, interest in establishing a TFCA at the confluence of the Limpopo and Shashe Rivers can be traced back to General Jan Smuts' decree for the establishment of the Dongola Botanical Reserve in 1922, which subsequently led to the creation of the Dongola National Park[46] in the 1940s (Limpopo/Shashe Transfrontier Conservation Area undated). The Dongola Wildlife Sanctuary Act No. 6 of 1947 that proclaimed some 190,000 ha as the area of the Dongola Botanical Reserve, was de-proclaimed by the National Party government in 1949 to allow for settlement by white farmers (Ryan 1999; Hall-Martin and Modise 2002; Tiley 2004).

Practical steps towards the creation of a national park for the promotion of a TFCA were taken in 1995 when the then Northern Province signed an agreement with the national Ministry of Environment and Tourism to proclaim the Vhembe–Dongola National Park, which comprised privately owned and state land. Subsequently, the Greefswald farm, which houses the famous Mapungubwe archaeological site, was incorporated into the park in 2000, a move that the South African government considered a breakthrough towards the establishment of the Limpopo/Shashe transfrontier park. SANParks together with WWF–South Africa, De Beers,[47] National Parks Trust and the PPF embarked on the process of purchasing farms to create the Vhembe–Dongola National Park that was proclaimed in 1998. When the park was renamed Mapungubwe and officially opened on Heritage Day,[48] 24 September 2004, the media reported the official opening of Mapungubwe National Park as that of the transfrontier park,[49] probably because Mapungubwe National Park is seen as central to the creation of the Limpopo/Shashe transfrontier park. While Botswana's Northern Tuli Game Reserve could be joined with Mapungubwe National Park to facilitate the creation of a TFCA, Zimbabwe has no protected area adjacent to its two neighbours. Zimbabwe's core area in the Limpopo/Shashe TFCA, the Tuli Circle Area, is state land on which hunting takes place.

The Final Draft of the Memorandum of Understanding[50] between the three countries did not specify which areas would be included in the TFCA, possibly because the areas that have been earmarked for that TFCA in Zimbabwe are highly contentious.[51] It is said that Botswana and Zimbabwe have agreed to the establishment of the Limpopo/Shashe TFCA, but Zimbabwe has been dragging its feet on the matter.[52] Notably, the final draft of the MoU for the Limpopo/Shashe TFCA has adopted the notion of stakeholders as in the Treaty of the Ais-Ais/Richtersveld. Article 3 of the final draft of the MoU upholds the rights of stakeholders as follows: (i) the Governments agree that the real rights recognisable under law of Stakeholders will be recognised and respected; (ii) a government will in terms of its domestic law enter into contractual arrangements with Stakeholders regarding the protection and regulation of matters affecting such rights; (iii) Stakeholders will become part of the process to establish the proposed Limpopo/Shashe TFCAs on a voluntary basis; (iv) the Governments acknowledge that in order to formalise the establishment of a TFCA, it will be necessary for each Government to take appropriate steps to designate or proclaim the land so included in terms of its relevant legislation and that such designation or proclamation will be preceded by the conclusion of a formal agreement between them acting in their sovereignty; and (v) it is recorded that certain land intended for inclusion in the TFCA is either owned or is subject to rights in favour of certain Stakeholders and that it will accordingly be necessary for binding agreements to be reached with Stakeholders prior to the aforementioned designation or proclamation of the TFCA.

The four examples of transfrontier parks illustrate the dominance of national parks in the creation of TFCAs in the region. Rather than creating TFCAs, the PPF-driven project of transfrontier conservation is preoccupied with expanding protected areas, particularly national parks. The PPF's focus on national parks is not new, but was firmly established by SANF as its frontrunner. In this sense, the establishment of transfrontier parks allows the founders of the PPF and its alliances to pursue their interest in national parks anew in an entirely different political environment. In the process, the rhetoric of TFCAs is useful to initiate plans for transfrontier parks, or to add a mosaic of land to transfrontier parks once such parks have been established. The preoccupation with, and interest in, transfrontier parks has led to the reopening of the discussion on how to transform TFCAs such as the Maloti–Drakensberg and the Lubombo into transfrontier parks. With regard to the Lubombo TFCA, preparations are underway to drop the fences in the Songimvelo–Malolotja and Usuthu–Tembe–Futhi by 2009[53] in order to recreate the migratory routes of elephants between Mozambique, South Africa and Swaziland. In practice, recreating migratory routes would imply the reclassification of areas in the TFCAs to allow for the proper management of the elephant population in the three countries. According to South Africa's tourism promotional magazine, *Discovery SA* (2006), Songimvelo is set to increase from 40,000 ha to 70,000 ha through the establishment of a transfrontier park.

Endnotes

1 Coincidentally, this is the same year in which the Club of Rome was formed.

2 Following World War I, Namibia (then South West Africa) was under the control of South Africa in accordance with the League of Nations mandate. Subsequently, South Africa exported its apartheid policies to Namibia, including the division of black people along ethnic lines. The Odendaal Commission Report of 1963 recommended the creation of 'homelands' for black people in Namibia. Against this backdrop, the Etosha National Park was dramatically reduced to gain land for Namibia's ethnic groups.

3 There is a well-established perception that post-apartheid South Africa dominates the subcontinent with respect to transfrontier conservation. While this is generally true, it should also be acknowledged that South Africa's hegemony was established before a post-apartheid political dispensation.

4 Those agencies were later to be deployed to promote the development of TFCAs.

5 Proof of this is in the original name of WWF, which was World Wildlife Fund for Nature. The name has since been changed to World Wide Fund for Nature to signify WWF's shift from overemphasis on wildlife towards a broad scope of nature conservation.

6 Most of the area of the former western Transvaal falls under the current North West Province.

7 The establishment of Pilanesburg Nature Reserve in 1979 was a practical step towards farming the area with wildlife.

8 This tripartite partnership continued after the demise of the apartheid state. For instance, the successor of SANF, the PPF also entered into agreement with the post-apartheid Department of Environmental and Tourism (DEAT) and the South African National Parks (SANParks) (Personal communication, Muleso Kharika, 2005; *Africa Geographic* 2005). The agreement facilitated the PPF's ambition to establish peace parks in southern Africa.

[9] It was the way by which the State President thanked the landowner on behalf of the nation.

[10] I will comment further on this aspect in the postscript in this volume.

[11] It has been estimated that the game farming and hunting industry expanded by 50% to 60% in the Makana and Ndlambe municipalities (Eastern Cape) between 2000 and 2004.

[12] The business sector financed nature conservation from their charity budget as part of their social responsibility.

[13] It is said that the Parks Board considered them as vermin.

[14] In the view of AECI.

[15] Relly was also Chairman of Anglo American at the time.

[16] Later in this Chapter I explain why the area around the Kruger National Park was earmarked as a site for the first transfrontier park in Africa, and why the first transfrontier park was actually established somewhere else.

[17] Mozambique was the best candidate for this, hence SANF planted the idea of transfrontier parks in that country.

[18] SANF regarded itself as the only international conservation body in Africa.

[19] His Cambridge PhD was based on the dynamics of elephants in Zambia's Luangwa Valley (*Africa Geographic* 2006).

[20] www.wild.org (accessed 3/5/2006).

[21] Player conceived the idea of the wilderness in 1957 (Player 1997).

[22] Anonymous informants.

[23] Abercrombie & Kent; ABSA Group; Anglo American Corporation of South Africa Limited; BOE/Repgro; De Beers Consolidated Mines Limited; Distillers Corporation Limited; FirstRand Limited; Forbes Group Limited; Iscor Limited; Nampak Limited; Nedcor; Sabi Sabi Private Game Reserve; Sanlam; Stellenbosch Farmers' Winery Limited; The South African Breweries Limited; The Standard Bank of South Africa; and Vodacom Group.

[24] Golden Tulip Hotels/Golden Tulip and The Lord Charles.

[25] AWP AG für Wirtschafts-Publikationen; Vendôme Luxury Group.

[26] Billiton PLC.

[27] The Edmond de Rothschild Foundation.

[28] It should be noted that Mlilwane is considered Swaziland's pioneer game reserve after the decimation of wildlife in that country during colonial rule.

[29] The enlargement of Swaziland's Mlilwane Game Reserve was one of the first three projects to be sponsored by SANF. The other two projects were the establishment of the 600 m long Engelhard Dam wall on the Letaba River and Eugene Marais Chair of Wildlife Management at the University of Pretoria (SANF 1993).

[30] The area was under the former bantustan of KaNgwane.

[31] www.environment.gov.za/projProg/TFCAs/lubombo.htm (accessed 8/5/2006).

[32] Correspondence, Wisdom Dlamini, 2005.

[33] These include a new road linking Hluhluwe to Maputo and improvements to the N2 between Mozambique, South Africa and Swaziland.

[34] Lubombo Ndumu–Tembe–Futi, Lubombo Nsubane–Pongolo, and the Lubombo Ponto do Ouro–Kosi Bay Marine and Coastal TFCAs.

[35] According to the Park Manager, M. Nkuebe, the South African Bob Phillip was responsible for fencing Sehlabathebe (Interview, 28/8/2005, Sehlabathebe).

[36] Anonymous interviewee, 29/8/2005, Sehlabathebe. By calling it a park, the Prime Minister could apply national park laws to arrest his opponents before they could come closer to where he lived – he was perceived as a dictator.

37 This does not mean that 1990 was a year in which co-operation between the two countries was initiated, as the Maloti–Drakensberg Programme had existed since 1982 (Sandwith 2003). Rather it is to acknowledge that 1990 was a crucial date in the development of the TFCAs as the idea of the TFCAs for southern Africa was in circulation and the political environment was undergoing a sea-change.

38 The zoning system in DAP included aspects of the proposed land use zones that appeared in the form of the 1976 Drakensberg Policy Statement (Town and Regional Planning Commission 2001).

39 It was initiated on the recommendations of Urban Econ, which was commissioned by the Town and Regional Planning Commission to investigate the need for, and feasibility of, a Regional Authority for the KwaZulu-Natal Drakensberg in 1996.

40 The workshop led to what has been termed the 'Giant Castle Declaration'.

41 Estimated to be 1 million per year.

42 Most TFCAs require the services of an international co-ordinator, who should facilitate the process of creating TFCAs. In the case of the Kgalagadi, such a co-ordinator was not necessary, because both the Wildlife Department and SANParks has already sealed the deal.

43 Interview, Sedia Modise, 25/8/2003, Gaborone.

44 A full discussion on the geographical complications surrounding the establishment of the GLTP is given in Chapter 6.

45 The initial name for this TFCA was Dongola/Vhembe.

46 This was later downgraded to a provincial nature reserve.

47 De Beers donated its 36,000 ha Venetia Limpopo Nature Reserve to Vhembe/Dongola National Park (Ryan 1999).

48 In South Africa.

49 *Getaway* (2004), *Mirror* (2004) and *Zoutpansberger* (2004).

50 Of 11 March 2002.

51 The strategic areas are communal land, such as the Maramani, who might have to be relocated in order to create a wilderness.

52 The MoU was not yet signed at the time of writing, i.e. May 2006.

53 allAfrica.com (accessed 24/5/2006).

Chapter 5

Southern African Development Community: a regional catalyst

Political and economic regionalism in Africa accompanied political changes on the continent from the 1960s onwards. Politically, the process emerged from the quest to reorganise the colonial state in line with new political demands, namely, the reunification of Africa and its people at a regional scale, a move that was championed by the former President of Tanzania, Julius Nyerere.[1] At the same time European powers were interested in assembling their former colonies into some sort of regional blocs or communities as in the case of the French Communaute Francophe and the British Commonwealth (Sidaway 2002). There were therefore competing regionalisms at the political level, with imperialism favouring regional projects, which were in line with metropolitan interests. The political dimension of regionalism is not unique to Africa, as regions elsewhere have developed as a response to other formations or as a strategy for political survival. For example, Japan was interested in the formation of the South-East Asia region, because the region offered that country a way out of a predicament in the 1930s (Sidaway 2002). The Soviet Union and the Eastern European communist republics established the Council for the Mutual Economic Assistance (CMEA) as a socialist alternative to the European Economic Community while APEC emerged as a response by Australia, New Zealand, the United States and Canada to a Pacific co-operation that threatened to exclude them (Pollard and Sidaway 2002; Sidaway 2002).

As in the formation of regional blocs world-wide, economic and political reasons for regionalism gave impetus to the division of Africa into the Arab Maghreb Union (UMA) in the north, the Common Market for Eastern and Southern African States (COMESA) in the south-east, the Economic Community of West African States (ECOWAS) in the west, the Economic Community of Central African States (ECCAS) as the central region and the Southern African Development Community (SADC) in the south. According to Adepoju (2001) the need to create regional groupings in Africa was reinforced by experiences of regional integration projects such as the Latin American Free Trade Association (LFTA), Caribbean Community (CAICOM), and the like. The existence of these regions in Africa begs the question of why only one region, southern Africa, was conducive for experimenting with TFCAs. The question is more pertinent if we consider the views that contemporary regional communities are 'a feature of the post-Cold War World' (Sidaway 2002, 5) and that many forms of current regionalism are manifestations of economic globalisation and prevailing forms of hegemony (Grant and Söderbaum 2003). In other words, regional integration is viewed as a collective effort by which states insert themselves into the global economy. As I have intimated in Chapter 1, TFCAs in southern Africa are seen as a manifestation of globalisation. If indeed TFCAs are a form of globalisation, what makes one region in the whole continent prone to that globalisation? My intention in this chapter is to account for the use of SADC as the launch pad for state-sanctioned TFCAs in Africa. Before we dwell on SADC as a conduit for TFCAs, some comments on the state of the environment in other regions of Africa are in order.

Africa's regional environmental outlook

The signing of the Treaty of the Southern African Development Community coincided with the global action plan for the environment and development that was adopted at the Rio Earth Summit in 1992. In the same year the CBD[2] was opened for signature. The Earth Summit fundamentally changed the conception of the environment by suggesting 'the conservation and sustainable use of biological diversity, and the fair and equitable sharing of benefits arising from its utilisation' (Secretariat of the Convention on Biological Diversity 2001, xvii). Accordingly, African states, like their counterparts elsewhere, had to use the CBD as the benchmark for national environmental policies. The CBD also became an important driver of environmental projects, more so because ratification of the CBD is a condition by which countries could access funds for environment-related programmes from the Global Environment Facility. The question that arises from the CBD is whether the regions in Africa responded differently to the CBD, and whether those responses account for the emergence of SADC as a leader in the development of TFCAs. Another related question is whether environmental conditions in SADC were significantly different from those in other parts of the African continent. A review of the status of signature, ratification, accession, acceptance and approval of the CBD in Africa reveals that the SADC region, with the exception of the island states of Mauritius and Seychelles, did not necessarily take the lead in ratifying the CBD (Table 5.1).

Table 5.1 Status of signature and ratification of the CBD in the SADC region.

State	Date of signature	Date of ratification
Angola	12 June 1992	1 April 1998
Botswana	8 August 1992	12 October 1995
Lesotho	11 June 1992	10 January 1995
Madagascar	8 June 1992	4 March 1996
Malawi	10 June 1992	2 February 1994
Mauritius	10 June 1992	4 September 1992
Mozambique	12 June 1992	25 August 1995
Namibia	12 June 1992	16 May 1997
Seychelles	10 June 1992	22 September 1992
South Africa	4 June 1993	2 November 1995
Swaziland	12 June 1992	9 November 1994
Tanzania	12 June 1992	8 March 1996
Zambia	11 June 1992	28 May 1993
Zimbabwe	12 June 1992	11 November 1994

Source: Adapted from the Secretariat of the Convention on Biological Diversity 2001, 260–262.

Most countries in other regions of Africa ratified the CBD before southern African states.[3] For example, in West Africa, Burkina Faso and Guinea ratified the CBD in 1993 and were joined by Cameroon, Gambia, Ghana, Nigeria and Senegal the following year, 1994. The same can be said about North and East Africa, where countries such as Tunisia (north) and Kenya (east) ratified the CBD in 1993 and 1994 respectively.

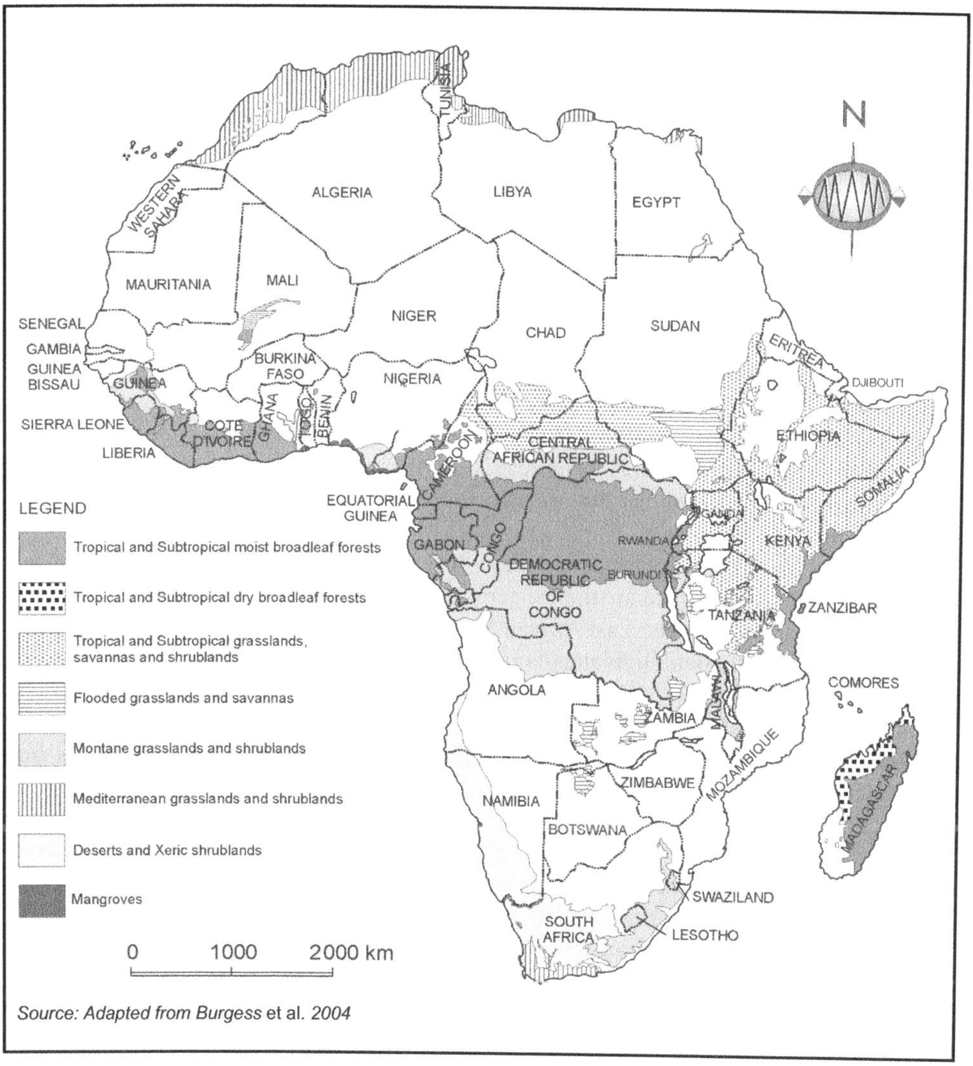

Figure 5.1 Africa's ecoregions.

The question of why other regions of Africa have been less enthusiastic about TFCAs demands that we should understand whether ecological conditions for cross-border collaboration existed in those regions, in the first place. It could be argued that Africa's political regions drew the attention of environmentalists, as each of the regions had something to contribute to biological diversity. For example, in northern Africa,[4] the dominance of arid and semi-arid environments has developed four cross-border hotspots: the Iamatong Mountains on the Sudan–Uganda border; the Jebel Marra volcanic massifs on the Chad–Sudan border; the Jebel Elba mountainous ecosystem on the Egypt–Sudan border; and the High Atlas Mountains between Morocco and Tunisia (UNEP 2002). The IUCN Commission on Protected Areas[5] considers North Africa and the Middle East[6] as some of the most diverse areas on earth, in part because those regions are located at a junction between three continents. In 2000 it was estimated that northern Africa had some

186 threatened species and 72 nationally protected areas (UNEP 2002). In West Africa, 'the upper Guinea forest, which extends from western Ghana through Côte d'Ivoire, Liberia and Guinea to southern Sierra Leone, is a biologically unique system that is considered one of the world's priority conservation areas because of its high endemism' (UNEP 2002, 83). The region is also home to the endangered Cross River gorilla, which is found in cross-border areas between Nigeria and Cameroon and is spread across Central and West Africa (UNEP 2002). In fact, the area stretching from south-western Cameroon through Equatorial Guinea, Gabon, Congo to the Democratic Republic of the Congo has been considered a cross-border wilderness area (Mittermeier *et al.* 1998; UNEP 2002). Similar cross-border environmental conditions of significance are found in Central Africa, which is home to the Congo basin. The basin has been described as the second largest contiguous area of forest in the world, and the largest in Africa (UNEP 2002). In eastern Africa, the Eastern Arc Mountain Forests, with more than 25% of endemic plant species, is another example of a cross-border ecosystem that meets the ecological conditions for transfrontier conservation areas outside SADC. Figure 5.1 confirms that Africa's ecoregions are not limited to southern Africa alone, but are also found in the northern, eastern, western and central regions of the continent.

Burgess *et al.* (2004) referred to Hilton-Taylor's work to suggest that the top ten imperilled ecoregions[7] in Africa are the: Eastern Arc Forests; Albertine Rift Montane Forests; Madagascar Subhumid Forests; East African Montaine Forests; Northern Congolian Forest-Savanna Mosaic; Cameroon Highlands Forests; Madagascar Humid Forests; Northeastern Congolian Lowveld Forests; Cross-Sanaga-Bioko Coastal Forests; and Somali Acacia-Commiphora Bushlands and Thickets. Equally, the richness and endemism index reveals that areas of globally and regionally outstanding status are not limited to TFCA sites, but are found in different parts of the continent (Figure 5.2). It has been estimated that there are 119 terrestrial ecoregions in Africa (Burgess *et al.* 2004), 28 of which are in sub-Saharan Africa (Brooks *et al.* 2001). Other areas such as the Serengeti/Maasai-Mara are considered natural cross-border areas, because they permit the seasonal migration of wildlife between Kenya and Tanzania. The former SADC senior official, Professor Paul Maro,[8] is of the view that Serengeti/Maasai Mara should have been the first TFCA in East Africa, because of wildlife migration and the absence of the fence between the two countries. It has been suggested that a cross-border protected area existed between colonial Burundi and Rwanda in the 1920s.[9]

In addition to those cross-border ecosystems, all politico-economic regions in Africa have contiguous national parks that could have been used to promote transfrontier conservation, and transfrontier parks in particular. The iconic species that have principally been used to promote the logic of ecological integrity for the establishment of TFCAs in southern Africa, the elephants, are spread across much of Central Africa, and a significant elephant population also exists in East and West Africa (Table 5.2). These examples have been used here to argue that various politico-economic regions in Africa meet the ecological requirements for establishing areas for transfrontier conservation, yet TFCAs that are couched in terms of the ecological integrity of ecosystems have not been a feature of those regions. In addition to ecological conditions, those regions also had the political environment that should have been supportive to TFCAs. My point here is that various parts of Africa meet the minimum political and ecological requirements for the development of TFCAs. Political and environmental considerations in East and West Africa could be used to substantiate this argument.

Table 5.2 African countries with a significant* elephant population.

Country	Estimated population
Benin	1,605
Botswana	121,866
Burkina Faso	3,090
Cameroon	14,242
Central African Republic	5,877
Chad	3,989
Congo	19,653
Côte d'Ivoire	1,064
Democratic Republic of Congo	49,558
Ethiopia	1,361
Gabon	76,857
Ghana	1,695
Kenya	27,732
Liberia	1,676
Malawi	2,296
Mozambique	21,502
Namibia	9,641
Tanzania	112,089
South Africa	14,926
Uganda	2,534
Zambia	20,323
Zimbabwe	89,128

Source: Adapted from Van Aarde and Jackson 2006

* More than 1000 elephants

Regionalism and TFCAs in East Africa

It would be naïve to suggest that other African regions lagged behind in the development of TFCAs, because they did not recognise the imperatives of regionalism or transfrontier conservation. On the contrary, regions such as East Africa experimented with the formation of a regional community, when Kenya, Tanzania and Uganda signed the Treaty on East African Co-operation on 1 December 1967, on the same day that the ill-fated East African Community (EAC) was inaugurated. It should be noted that the EAC was the outcome of attempts by the independent East African states to create the East African Federation,[10] as a practical step towards the realisation of Pan-Africanism. Of relevance to our discussion is the fact that the EAC aimed to promote co-operation in many areas, including wildlife conservation and natural resources. Though the EAC collapsed in 1977, its formation demonstrates that the spirit of regional co-operation prevailed in other parts of Africa long before the formation of SADC in 1992. The Intergovernmental Authority on Drought and Development (IGADD)[11] that was formed in 1986 carried the process of regionalism in East Africa forward. In the mid-1990s IGADD was revitalised into a fully-fledged regional entity that assumed the same status as that of SADC. It moved away from its focus on drought and desertification to encompass the broad mandate of development, hence it was renamed the

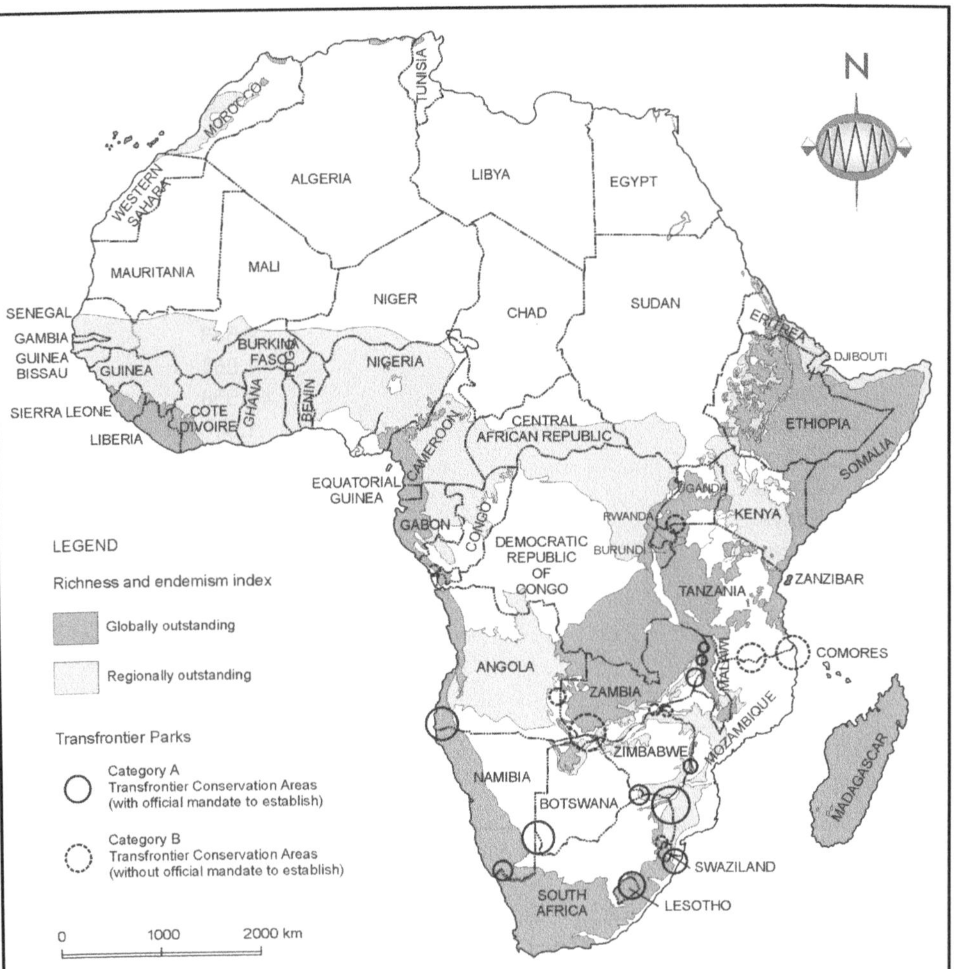

Source: Adapted and modified from Burgess et al. *2004*

Figure 5.2 Biodiversity indices and TFCA sites.

Intergovernmental Authority on Development (IGAD) in March 1996.[12] It should be noted that one of the first three priority areas of co-operation of the newly formed IGAD was food security and environmental protection.[13]

In the area of natural resources and the environment, IGAD aims to harmonise policies, initiate and promote programmes and projects that would ensure sustainable development of natural resources and environmental protection. These aims are in line with those of the EAC Development Strategy (1997–2000), which encourages co-operation in environmental and natural resources management and tourism and wildlife management (East African Community Secretariat 2001).[14] Despite these developments, TFCAs remained a rare feature in East Africa. Research on conservation in that region appears less concerned with the idea of TFCAs. For example, contributors to Newmark's (1991) edited volume, *The Conservation of Mount Kilimanjaro*, endorsed eight priorities[15] for the conservation of Mount Kilimanjaro, but none of their priorities suggest the creation of TFCAs. This is rather surprising given that the authors emphasised the need for corridors, and their work was published by the IUCN.[16] Newmark (1991, 112) concluded that, 'in order to ensure that the existing wildlife corridor on the Kenyan side of the Tanzanian–Kenyan border is maintained, the Wildlife Division and Tanzania National Parks should establish a dialogue with the Kenyan authorities'. Equally, Brockington's (2006) work in Tanzania emphasises the ways in which environmentalism in that country is pursued at the national, local and village levels, but is silent on environmentalism at the supranational level.

Nevertheless, there are signs that the TFCA idea is beginning to spread to East Africa. For example, the EAC Development Strategy 2001–2005 includes the intention to harmonise management programmes for transboundary ecosystems. According to the East African Community Secretariat (2001), priority will be given to the major watershed/catchment areas of Mt Elgon, Mt Kilimanjaro, Ewaso Ngiro and the Pemba Channel. However, the EAC Development Strategy is vague on whether co-operation in wildlife would be in the form of transfrontier parks. Instead, it points to the need for the study that should be undertaken to investigate co-operation in hunting, anti-poaching, preservation of wildlife migratory routes and conservation of fauna.[17] Undoubtedly, GEF will assist the region to create cross-border protected areas. Indeed, such a move has come in the form of the East Africa Cross Border Project, which has been implemented in partnership with the IUCN–East Africa Regional Office and the African Centre for Technology Studies. Kenya and Tanzania are key to the development of cross-border protected areas in East Africa. For example, the two countries are involved in three of the four East Africa Cross Border Project sites. These are: Bukoba to Rakai/Mbarara on the Tanzania–Uganda border; Monduli to Kajiado on the Tanzania–Kenya border; Same to Taita–Taveta on the Tanzania–Kenya border; and Turkana to Moroto/Kotido on the Kenya–Uganda border. In the view of Ngoile (in Shechambo *et al.* 2001), the East Africa Cross Border Project is a humble beginning in the development of a regional strategy by which cross-border environmental issues will be resolved.

Recent developments in the Selous–Niassa Corridor and the Central Albertine Rift point to the beginning of the era of transfrontier parks in East Africa. Though the idea of establishing the Selous–Niassa Corridor can be traced back to proposals that the Selous Game Reserve[18] required an emergency programme because of the imminent danger of losing its rhino and elephant (Baldus *et al.*, 2003), the current development points to the process of connecting the world's largest elephant ranges to the Lake Malawi/Niassa/Nyasa ecoregion, which will be a springboard for launching the TFCA between Malawi, Mozambique and Tanzania. Baldus *et al.* (2003, 60) conceded that 'after 15 years of effort by different partners to conserve one of Africa's greatest wilderness and wildlife areas, the time might be ripe to begin a new chapter, the one of international partnerships'. In essence, the 'new chapter' referred to is the development of a TFCA in which the Selous Game Reserve (Tanzania) would be joined together with Niassa Game Reserve in Mozambique. The German Government funded a 4-year (2000–2003) research project that supported the need for the Selous–Niassa Corridor, arguing that the corridor is not only a transit route for elephants between the two game reserves in the north and south, but also sustains its own sizable resident population (Baldus and Hahn forthcoming). In 2006, the German Development Bank committed EUR 5 million to the development of the Selous–Niassa Corridor on the Tanzanian side. It should be emphasised that the Selous–Niassa Corridor is central to the development of the Lake Malawi/Nyasa/Niassa TFCA. George Mkondiwa (2003) of the Secretariat for Natural Resources and Environmental Affairs of Malawi, estimated that 1.6 million people from the riparian states of Malawi, Mozambique and Tanzania rely on the fisheries resources of that

TFCA. There are indications that that TFCA will expand to Zambia through the development of the Mtwara Development Corridor for which the four countries signed an agreement on 15 December 2004. The Mtwara Development Corridor provides the necessary infrastructure for tourism development in the region. Similar developments have taken place in the Central Albertine Rift.

In the Central Albertine Rift the Virunga Volcano massif and Bwindi Impenetrable Forests are considered the only habitat of the mountain gorilla in the world. The International Gorilla Conservation Programme (IGCP)[19] has promoted a regional framework for the conservation of the gorilla in East Africa since 1991 (Rainer *et al.* 2003). The framework operates in the form of regional meetings, joint patrols, and cross-border visits while, at same time, supporting protected area authorities of relevant countries. The aim is to lay the foundation for a regional collaborative structure. The IGCP has promoted the need for regional collaboration on ecological, economic and political grounds, the ultimate goal being to establish a TFCA. Clearly, the political instability and violent conflict that engulfed the Great Lakes Region after the collapse of the Mobutu regime in Zaïre in 1993 and genocide in Rwanda in 1994 complicated efforts towards establishing a regional conservation framework. Nonetheless, the IGCP achieved the first step towards establishing the TFCA through the signing of the MoU on the Collaborative Conservation of the Central Albertine Rift Transfrontier Protected Area Network (CARTPAN) by Rwanda, Uganda and the Democratic Republic of Congo in 2004. The three countries agreed to jointly protect Volcano, Mgahinga Gorilla, Bwindi Impenetrable, Queen Elizabeth, Semliki, Ruwenzori Mountains, Kibale and Virunga National Parks as a transfrontier ecosystem (Trilateral Memorandum of Understanding 2004). The three countries signed the Tripartite Declaration on the Transboundary Natural Resources Management of the Transfrontier Protected Area Network of the Central Albertine Rift on 14 October 2005. The intention is to take this forward in the form of developing a collaborative protocol among the governments of the three countries.

Four inferences can be made from the CARTPAN initiative. Firstly, the CARTPAN is based on the model of a transfrontier park, in which national parks are used as building blocks (see Chapter 3). Secondly, although the idea of a peace park has been played down by the designation of the initiative as a 'transfrontier protected area network', the language of peace parks is evident in the MoU and in the writings of those supporting the initiative such as Rainer *et al.* (2003). Article 3 of the MoU states that the transfrontier objectives are, among other things, 'building trust, understanding and co-operation among wildlife authorities, nongovernmental organisations, communities, users and other stakeholders to achieve sustainable conservation and thereby contribute to *peace*' (Trilateral Memorandum of Understanding 2004, 3). The emphasis on wildlife is relevant to my third inference, namely, the intention to use the protected area network to promote ecotourism. In fact, two[20] of the signatories of the Tripartite Declaration were ministers responsible for tourism in their respective countries. According to Rainer *et al.* (2003), Rwanda and the former Zaïre established gorilla-based tourism before the formation of the IGCP in 1979. Against this backdrop, the IGCP used the existing tourism industry to promote the view that a TFCA will lead to the expansion of that industry. Fourthly, the initiative is also founded on the belief that the CARTPAN will contribute to socio-economic development, especially poverty alleviation. In support of the socio-economic rationale for the TFCA between the two countries, Rainer *et al.* (2003, 197) argued that, 'very few alternatives outside of subsistence farming exist for income generation. Tourism, or more specifically ecotourism, can provide one of the more viable options to the surrounding population in addition to the much-needed funds for the [protected areas]'.

Regionalism and TFCAs in West Africa

As in East Africa, West African states also embraced regionalism, but the ideas of transfrontier conservation did not readily translate into transfrontier parks in that region. Sixteen West African states[21] signed the Treaty to establish the Economic Community of West African States in Lagos on 28 May 1975. Chapter VI of the Treaty focuses on co-operation in environment and natural resources, and in Article 29 of that chapter, member states undertook to protect, preserve and enhance the natural environment of the region. This was to be achieved by adopting policies, strategies and programmes at national and regional levels and by establishing appropriate institutions (ECOWAS 1975). Nonetheless, ECOWAS emphasised co-operation on mineral and water resources, as per Article 31 of Chapter VI, by which member states undertook to co-ordinate their programmes for the development and utilisation of mineral and water resources and to acquire appropriate skills and technologies to realise that goal. Accordingly, ECOWAS initiated the Integrated Water Resources Management following the Ouagadougou Conference in 1998.[22] The Ouagadougou Statement urged the governments of member states to, among other things, create the frameworks for regional co-operation on integrated water resources management and for joint management of shared basins among riparian states. The Water Resource Co-ordination Unit was created to assist member states to achieve those goals.

Other institutions that are supportive of the idea of governing shared natural resources in West Africa include the Mano River Union and the Niger River Basin Authority. Whether these developments can be regarded as a manifestation of transfrontier conservation areas in West Africa is debatable, more so because the notion of cross-border areas in that region is not necessarily couched in terms of nature conservation areas, but is more focused on the socio-economic component of regionalism. Regionalism in that region is also premised on the view that colonial borders obstruct regional integration, hence the need to define borders as zones of contact rather than separation. For such contact to materialise, border zones should be created to integrate neighbouring communities across the national boundaries of contiguous states. Accordingly, the Sikasso consensus[23] defines a cross-border area as 'a geographical area that overlaps between two or more neighbouring states, and whose populations are linked by socio-economic and cultural bonds' (Diarrah 2002, 2).

While the concept of a cross-border area (in the context of West Africa) is concerned with the legacy of colonial borders that underpin the discourse of TFCAs in southern Africa, its focus is on the regional integration process. Its specific objectives are to: create homogeneous areas of development and co-operation; encourage trade and social and cultural exchanges; create better conditions for the circulation of goods and people; promote cross-border co-operation and help communities undertake common development actions in the fields of health, education, arts, sport and culture, agriculture, transport, energy, environment and industry; harmonise national sectoral policies on health, education, infrastructure, sports, arts and culture; and intensify trade flows by providing infrastructure and capital equipment for border markets and by creating and strengthening production-to-distribution chains (Diarrah 2002). Unlike the demarcation of ecoregions, the Sikasso consensus proposed that the cross-border areas in ECOWAS should be demarcated on the basis of existing national or local administrative boundaries, and that states should jointly demarcate areas. Arguably, the proposed anthropogenic cross-border areas in West Africa emerged as part of the policy instrument to enhance regional integration and to support the implementation of the progressive migration policy of that region. As I have argued in Chapter 5 and elsewhere

(Ramutsindela 2004b), the designation of cross-border areas in the form of transfrontier parks in southern Africa favours the migration of wildlife rather than that of the local residents of neighbouring countries.

It should be noted that ECOWAS also considered tourism as a critical area of co-operation among member states in terms of Chapter VII, Article 34 of the Treaty of ECOWAS. Member states undertook to strengthen regional co-operation in tourism through the: promotion of intra-community tourism; harmonisation and co-ordination of tourism development policies, plans and programmes; harmonisation of regulations governing tourism and hotel management activities; institutionalising a community reference framework for tourism statistics; and by jointly promoting tourism products portraying the natural and socio-cultural values of the region (ECOWAS 1975). In this context, the vision for tourism development was not tied to that of TFCAs, in which tourism is promoted through the establishment of a cross-border wilderness.

The experiences in East and West Africa referred to above show that the attempts at co-operation over the management and use of natural resources that occurred in those regions did not necessarily translate into TFCAs. The same can be said about the Nile River, which has a long history of co-operation over its shared water resources. Despite that long-established political tradition, trans-border conservation in North Africa is a recent phenomenon. This is so though more than 30 regional meetings were convened between 1993 and 1999 to promote inter-Arab co-operation on biodiversity conservation (UNEP 2002). Algeria and Niger alluded to the prospect for co-operation between West and North Africa in protecting the Saharo–Sahelian fauna during their joint declaration of adopting the CBD (Secretariat of the Convention on Biological Diversity 2001).

One of the axial themes in transfrontier parks relates to the act of 'breaking down colonial boundaries'. This begs the question of why environmentalists have been slow in breaking down colonial boundaries in Africa's regions other than southern Africa. What is so colonial about boundaries in southern Africa as compared to other parts of Africa? It is a historical fact that Africa as a whole suffered from the imposition of colonial boundaries (Asiwaju 1985; Gakwandi 1996). Geographically, West Africa is the most colonially fragmented region in Africa. It was on the basis of that fragmentation that Kwame Nkrumah (1961, xi) mourned and counselled after independence that, 'it is heartening to see so many flags hoisted in place of old; it is disturbing to see so many countries of varying sizes and at different levels of development, weak and, in some cases helpless. If this terrible state of *fragmentation* is allowed to continue, it may well be disastrous for us all'. The point here is that West Africa was equally an ideal candidate for transfrontier projects, not least because of the intensity of the geographical fragmentation of states in that region. Equally, East Africa was far ahead of southern Africa in matters relating to the integration of people in that region. For instance, it promoted Swahili as a regional language and established a federal University of East Africa with university colleges in Kenya, Tanzania and Uganda. Though some of the regional initiatives collapsed on the way, they do however demonstrate the fact that East Africa had, and still has, the political will towards integration.

It could also be argued that the notion of peace parks is more compelling in Africa's north, west, and east regions where violent conflict has been long and intense than in southern Africa. Interestingly, discussions on cross-border protected areas and the broader regional integration process in East and West Africa are also couched in the same anti-colonial language that has been used to support the creation of TFCAs in southern Africa, yet, transfrontier parks remained alien to those regions. Against this backdrop, the uniqueness of SADC with regard to transfrontier parks can more appropriately be understood by carefully analysing the opportunities for TFCAs that emanated from the creation of SADC and how that regional entity operated. As we have noted in Chapter 4, transfrontier parks were initiated outside the structures of SADC by non-governmental organisations and therefore did not originate as a regional initiative. This begs the question of how the SADC region became a vehicle through which the idea of transfrontier parks could be pursued. Is the process of regionalism in SADC unique to the continent? Are southern African states exceptional in their handling of cross-border environmental issues? Answers to these questions demand an understanding of regionalism in southern Africa and the links between that process and the environmental agendas.

The rise and working of SADC

Globalisation theorists have suggested that regional integration is a manifestation of, and response to, the interplay of global forces (Cleaver 1997; Cheru 2002; Sidaway 2002). In particular, the liberalisation of economies and the weakening of the state are said to have created the need for the rearrangement of states into regional blocs. A closer inspection of SADC reveals that the region was more a response to regional circumstances that affected the fortunes of the states in that region (Gibb 1996; Simon 1998) than a preoccupation with the global political and economic environment. In no way do I suggest that there were no global interests and rivalries in the region. Evidently, Cold War politics was played out in the region, with South Africa serving western interests and forming a buffer zone against the spread of communism in the region. Rather, I suggest that the formation of a regional bloc in southern Africa was not a direct response to the bi-polar Cold War politics.[24] SADC owes its origin to the formation of the Southern African Development Co-ordination Conference (SADCC), which was established in 1980 to co-ordinate the economies of nine independent countries[25] and to reduce their dependence on apartheid South Africa. The SADCC's Programme of Action[26] included: the creation of a Southern African Transport and Communications Commission based in Maputo; measures to control foot and mouth disease in cattle throughout the region; the preparation of a food security plan for the region; the establishment of a Regional Agricultural Research Centre specialising in drought-prone areas; plans for harmonisation of industrial and energy policies; sharing of national training facilities within the region; and conducting studies leading to proposals for the establishment of a Southern African Development Fund (SADCC 1980). Rather than creating a bureaucracy that would carry out the Programme of Action, the SADCC Summit resolved in July 1981 that the governments of member states were responsible for implementing that programme (SADCC 1981).[27]

Of significance to the discussion of this chapter, and the volume as a whole, is that the SADCC Council of Ministers allocated Malawi the responsibility for co-ordinating a regional approach to the development of fisheries and wildlife (SADCC 1981). Malawi therefore had the opportunity to influence regional policies on wildlife and should have taken the lead in the development of TFCAs in the region. However, that leadership was taken over by South Africa, which was not even a member of the SADCC, because of the influence that the South African-based SANF had on nature conservation (see Chapter 4). Moreover, Malawi's nature conservation policies focused on the forestry and fisheries and were also inward-looking, as I will be showing shortly. The SADCC also had a disjointed 'policy' on the protection of biodiversity as evidenced by the allocation of sector responsibilities. For example, according to the decision of the SADCC Council of Ministers, Lesotho was entrusted with the responsibility of co-ordinating regional activities related to Soil and Water Conservation and Land Utilisation (SWCLU), while Zimbabwe was responsible for regional activities related to food, agriculture and natural resources (SADC 1996a).

It follows that the structure of the SADCC focused on the political agenda of liberation, a situation that gave room for non-state actors to attend the SADCC Summits. For example, members of the African National Congress (Oliver Tambo), the Pan Africanist Congress (John Pokela) and a representative of SWAPO attended the fifth SADCC Summit in Gaborone in July 1984 (SADCC 1984). This was not a once-off attendance, as members of the liberation movements continued to attend the SADCC Summits from 1984 onwards.[28] Clearly, the participation of the liberation movements in SADCC Summits was important for developing a common political position towards South Africa and its regional destabilisation policy. Unsurprisingly, SADCC allowed members of the liberation movements to attend its Summits after the organisation had raised the concern in 1983 that South Africa had stepped up its aggressive policy against member states (SADCC 1983). The attempts by the SADCC member states to develop a common policy towards apartheid South Africa were frustrated by the reliance of those member states on South Africa's infrastructure for trade; the lack of intra-trade among the member states; and the multiple memberships[29] that the members had. Admittedly, external pressure also weakened the development of SADCC as a unified regional bloc. For example, the US Congress passed the legislation to assist SADCC in 1987, but excluded Angola and Mozambique from such programmes.

Transforming SADCC into a regional community

The need to formalise SADCC was suggested at the SADCC Summit in Harare in August 1989. At that Summit, Council members were instructed to 'formalise SADCC, and give it an appropriate legal status, taking into account the need to replace the SADCC Memorandum of Understanding with an Agreement, Charter or Treaty to be prepared in readiness for signature for the 1990 Summit'[30] (SADCC 1989, 2). There is a strong view that the end of the Cold War in 1989 accounts for the restructuring of the SADCC. Noting that the region was undergoing fundamental economic, political and social changes, the SADCC Summit directed Council in 1991 to 'undertake the necessary work to identify options and strategies for regional co-operation in a post-apartheid era' (SADCC 1991, 3). Subsequently, the SADCC Summit approved the formalisation of SADCC in August 1992, leading to the formation of a new organisation, the Southern African Development Community (SADC). According to Article 5 of the Treaty of the Southern African Development Community, SADC had eight objectives of which four were to have a direct bearing on transfrontier conservation. These are to achieve development and economic growth, alleviate poverty, enhance the standard and quality of life of the people of southern Africa and support the socially disadvantaged through regional integration; promote and defend peace and security; achieve sustainable utilisation of natural resources and effective protection of the environment; and to strengthen and consolidate the long standing historical, social and cultural affinities and links among the people of the Region (SADC 1992).[31] In addition, member states of SADC agreed to co-operate in areas of, among other things, natural resources and the environment.[32] Though SADC member states recognised the need for co-operation, the manner in which that co-operation would be implemented in the management of natural resources was vague, as evidenced by the absence of the relevant protocols at the time (i.e. 1992). The Declaration and Treaty of the Southern African Development Community suggested that 'policy measures [were to] be taken, and mechanisms instituted to protect the environment, and manage natural resource utilisation with a view to achieving optimum sustainable benefits for the present and future generations of Southern Africa' (SADC 1992, 6–7).

The transformation of SADCC into SADC was accompanied by shifts in environmental policies. The SADC policy and strategy for sustainable development that was developed in 1993–1995 was an attempt to move 'away from the past approach of managing the environment in fragmented, sector-oriented manner' (SADC 1996a, 6). The implication of the transformation was also that 'projects' were to be realigned to the imperatives of the newly formed regional community. A year after the formation of SADC (i.e. 1993) the SADC Summit placed a moratorium on new projects (SADC 1993), and in 1994[33] the primary focus of the SADC organs was the building of SADC as a community. In practice this meant that the SADC Programme was streamlined to 'ensure that only projects with a regional impact and that can be viable and self-sustaining remain in the programme' (SADC 1994, 4). In its efforts to give clarity to the spatial limits of SADC, the SADC Summit decided on splitting the Preferential Trade Area (PTA) into the Southern (SADC) and Northern Regions (Non-SADC) in 1994. Effectively, the decision was meant to separate SADC from COMESA, with far-reaching implications on the implementation of protocols. That is to say that protocols, as legal instruments, determined the ways in which the member states of SADC would co-operate as members of a distinct regional community.

In contrast to the populist view that TFCAs would encourage co-operation among SADC members, the SADC Protocol on Shared Watercourse Systems that was signed on 28 August 1995,[34] was the first protocol to put such co-operation into practice in the region. The General Principles of that Protocol as per Article 2 reveal that member states were more concerned with the management and sharing of water resources than designating areas of protection in the form of TFCAs. This is not to suggest that the Protocol did not consider the protection of shared water resources as an issue. Rather, it is to argue that the concern with the protection of watercourse systems was not tied to the idea of TFCAs.[35] For example, Sub-Section 12 of Article 2 of the Protocol alludes to the fact that 'Member States shall maintain and protect shared watercourse systems and related installations, facilities and other works in order to prevent pollution or environmental degradation' (SADC 1995a, 6), but the objectives of River Basin Management Institutions are not to establish TFCAs. Instead, they are meant to: develop a monitoring policy for shared watercourse systems; promote the equitable utilisation of shared watercourse systems; formulate strategies for the development of shared watercourse systems; and monitor the execution of integrated water resource development plans in shared watercourse systems (SADC 1995a).[36] The objective of the Protocol as a whole is to 'enhance co-operation in the optimal utilisation and conservation of shared watercourse systems in the SADC region' (SADC 1995b, 52). Arguably, proponents of TFCAs did not promote the spirit of co-operation and the management of natural resources expressed in that Protocol, because they were more interested in wildlife than water resources. In other words, co-operation over shared water resources was not instrumental towards the creation of TFCAs, because water resources neither provided ready-made building blocks for protected areas[37] nor were they relevant to the envisaged tourism industry.

The Sector that should have spearheaded the development of transfrontier parks in the region, was Soil and Water Conservation and Land Utilisation, renamed Environment and Land Management Sector (ELMS), which later made a new environmental policy and strategy for SADC. According to Maro,[38] ELMS emerged from attempts to develop the SADC statement for the 1992 Rio Summit. The existing Sector on Soil and Water Conservation was considered too specific and inadequate, hence there was a need to design a broad concept of the environment that would be relevant for the whole of SADC. It was maintained that 'the new SADC policy [for environment and resource management] must be developed and implemented as an integral part of a larger SADC agenda and strategy for equity-led growth and sustainable development in and among the countries of the SADC region' (SADC 1996b, ix). The new policy broadened the scope of the SWCLU to protect and improve the health, environment and livelihoods of the people in southern Africa; preserve the natural heritage, biodiversity and life-supporting ecosystems in southern Africa; and support regional economic development on an equitable and sustainable basis for the benefit of present and future generations (SADC 1996a, b). The change from SWCLU to ELMS was part of the centralisation tendency that had come to characterise the restructuring of SADC.[39] The process of centralisation was promoted on the basis that there was a duplication of responsibilities and that Sectors were too donor-dependent.[40] In practice, Lesotho remained a nominal host of ELMS whereas policies were directed from the Secretariat in Gaborone, Botswana. The same can be said about Malawi, which remained responsible for Inland Fisheries, Forestry and Wildlife Sector.

It could be argued that despite changes in SADC environmental policy and strategy, the government of Malawi was less enthusiastic about promoting cross-border protected areas as part of SADC environmental policy. A closer inspection of Malawi reveals that its national environmental policies and actions were more a response to the Rio Earth Summit and the Johannesburg World Summit on Sustainable Development. For example, the government adopted the National Environmental Action Plan in 1994, which was supported by the National Environmental Policy of 1996. The Action Plan was developed immediately after the Rio Earth Summit. In addition, the core of Malawi's National Strategy for Sustainable Development is based on the nine thematic areas of the Johannesburg Plan of Action (Malawi 2004).

Of relevance to the discussion in this chapter is the theme on Protecting and Managing the Natural Resource Base of Economic and Social Development. With 88 gazetted forest reserves covering some 28% of the country and fish providing 60–70% of the total animal protein in the country, it is understandable that Malawi's nature conservation policy focused on forestry and fisheries. In the light of the discussion in Chapters 2 and 3, forestry and fisheries were marginal to the propositions for peace parks in Africa. The point here is not that transfrontier parks cannot be established on the basis of forestry and fisheries, as indeed there are transfrontier parks of that nature. In East Africa, Coulter and Mubamba (1993) proposed that extending four existing wildlife land parks into the lake to form underwater parks is one of the appropriate strategies for conserving Lake Tanganyika. In their view, the boundaries of the underwater park would be determined on an ecological basis to encompass the entire local distribution of species. Despite the possibility of establishing transfrontier parks for marine resources, East Africa, and Malawi in particular, was not 'in the game of transfrontier parks'.[41] Equally, the opportunity of using national parks and wildlife to promote the creation of transfrontier parks between Malawi and its neighbours was not realised, mainly because transfrontier parks did not feature in that country's National Strategy for Sustainable Development. Instead, Malawi's goal with regard to national parks and wildlife was to increase law enforcement capacity in protected areas with adequate and well-trained and equipped staff, increase community-based natural resource management, improve knowledge of wildlife resources through appropriate wildlife research, control human and wildlife conflicts, improve the institutional structure of the Department of National Parks and Wildlife, initiate training needs for the sector and to reward staff properly (Malawi 2004).

It should be pointed out that SADC, through its Sectoral office in Malawi, opted to establish and use the Biodiversity Focal Point (BFP),[42] which acted as a 'clearing house for projects, programmes and activities pertaining to biodiversity in SADC' (Nyasulu 1997, 9).[43] The intention was neither to substitute various SADC sectors nor to replace regional bodies such as the IUCN. Rather, the aim was to use the BFP to collate information about conservation activities and research results in order to formulate a SADC position on aspects of the Convention of Biological Diversity. These included facilitating the development of a regional biodiversity strategy and action plan, and promoting the preparation of comprehensive biodiversity maps and management plans for each member state. These expectations were unrealistic given that the BFP had only two staff of a professional level, a technical advisor and finance assistant (Nyasulu 1997).

The process of rationalising SADC, which was recommended by the Main Report on the Review and Rationalisation of the SADC Programme of Action in April 1997, sheds some light on regional dynamics and their implication for the development of TFCAs in the region. The Review focused on the strategy and framework for building the SADC community and on the policies, priorities and strategies of all 16 SADC Sectors, the aim being to recommend amendments to the SADC portfolio of policies and strategies. The main finding of the Review was that, although member states responsible for particular sectors have developed a sense of 'ownership', the number of Sectors and Sub-Sectors was unsustainable, more so because Sectors were allocated for political reasons rather than as a result of the comparative advantage of the member states responsible for the Sector. Besides, Sectoral responsibilities were fragmented and intermixed with policy and institutional issues (SADC 1997). Under these circumstances two options were suggested for the restructuring of SADC. The first option was to rationalise the status quo by redefining the Sectors and merging functions and to allocate one sector per member state. The Review suggested that there should be 12 Sectors covering: agriculture;[44] transport, communications and information technology; industry and trade, finance and investment; human resources development; energy, water, environment, health and population; tourism; culture and mining (SADC 1997). The second option was to create clusters that would promote cross-sectoral economic integration. The Review invoked the seven areas of co-operation identified in Article 21(3) of the SADC Treaty to suggest the following seven clusters: food security, land and agriculture; infrastructure and services; industry, trade, investment and finance; human resources development, science and technology; natural resources and the environment; social welfare, information and culture; and politics, diplomacy, international relations, peace and security (SADC 1997). The Review recommended the second option on the basis that clustering areas of co-operation would lead to the development of region-wide programmes and structures. The attempt to rationalise the SADC Sectors created confusion in the region, mainly because the tradition of working through Sectors is deeply entrenched in the region while, at the same time, donors were increasingly becoming interested in promoting regional programmes that did not necessarily fit into the Sectors. The implication of the last point is that donors could launch their regional programmes from a member state, which did not have the traditional Sector responsibilities. This is particularly relevant for South Africa, which has been viewed by analysts as the force behind TFCAs, yet that country's Sector responsibilities are Finance and Investment and Health rather than the environment (Table 5.3).

Table 5.3 SADC sectoral responsibilities.

Member state	Responsibilities
Angola	Energy
Botswana	Agricultural research, Livestock Production and Animal Disease Control
DRC	N/A
Lesotho	Water, Environment and Land Management
Malawi	Inland Fisheries, Forestry, Wildlife
Mauritius	Tourism
Mozambique	Culture, Information, Sport, Transport, Communications and Technology
Namibia	Marine Fisheries and Resources, Legal Affairs
Seychelles	N/A
South Africa	Finance and Investment, Health
Swaziland	Human Resources Development
Tanzania	Industry and Trade
Zambia	Mining, Employment and Labour
Zimbabwe	Food, Agriculture and Natural Resources, Crop Production

Source: SADC 1999a

When the SADC Council of Ministers reaffirmed the position of the Government of Malawi as responsible for providing leadership in conservation and sustainable use of biological resources in the SADC region in September 1997, the PPF had already mapped out seven TFCA sites in South Africa and its neighbours.[45] It should be noted that though Malawi was responsible for wildlife for the SADC region, the PPF excluded that country from its initial proposals for TFCAs in 1997. This was the case despite the involvement of SANF in nature conservation in Malawi.[46] The PPF focused on South Africa and its neighbours, most probably because it wanted to capitalise on the opportunities that South Africa offered following the demise of the apartheid state and the end of civil war in Mozambique. Subsequent to the identification of TFCAs in South Africa and its neighbours, the PPF also showed interest in developing a TFCA in Malawi (PPF 1998), and by 1999, the GEF Lake Malawi/Nyasa Biodiversity Conservation Project had been established in Malawi and Mozambique. The point here is that the PPF was instrumental in pushing TFCAs in southern Africa though it operated outside the formal structures of SADC. In other words, TFCAs were not started by the SADC Sectors, but the restructuring of those Sectors allowed the private sector and NGOs to push the TFCA's agenda through countries, such as South Africa and Mozambique, that had no Sector responsibilities for wildlife. This was possible since the Sectors had been unbundled and weakened.

Evidently, the launching of the GEF-funded Southern African Biodiversity Support Programme (SABSP) in Lilongwe in September 2000 contributed to the regionalisation of TFCAs, as the Programme adopted an ecosystems approach. For instance, it highlighted the significance of conserving the following ecosystems in southern Africa: the arid- and semi-arid ecosystems (Karoo–Kalahari–Namibia region); the whole of the Mediterranean-type ecosystem (Cape Floristic kingdom); coastal, marine and freshwater ecosystems (Zambezi, Okavango delta and Kafue wetlands); forest ecosystems (Guineo–Congolian forests of Angola, the Usambara/Inhambane forests of Mozambique, the Afro-montane forests of Angola, Malawi, Mozambique, South Africa, Swaziland, Zambia and Zimbabwe); and mountain ecosystems such as the Huambo and Huila (Angola), the Chimanimani (Mozambique and Zimbabwe), the Nyika (Malawi and Zambia), the Drakensberg (South Africa and Lesotho) and Mount Mulanje (Malawi) (SADC 2002). In line with the ideology of TFCAs, SABSP promoted the view that ecosystems in southern Africa exist within transboundary areas and therefore required regional co-operation.

It should be noted that SABSP is a regional arm of the Biodiversity Support Programme, which is a USAID-funded consortium of WWF, The Nature Conservancy and the World Resource Institute. SABSP is described as

> a capacity building programme that has the main goal of promoting the conservation and sustainable use of biological diversity in southern Africa by strengthening regional biodiversity planning, co-operation and information exchange ... by strengthening regional institutional structures and assisting in the establishment of a sustainable financing mechanism of these. This will lead into improvements in both national and transboundary biodiversity and ecosystem management and conservation ...' (SADC 2002, 13).

SABSP capitalised on the SADC Protocol on Wildlife Conservation and Law Enforcement that was signed by member states on 18 August 1999. The Protocol recognised the importance of creating the transfrontier conservation area, defining it as 'the area or component of a large ecological region that straddles the boundaries of two or more countries, encompassing one or more protected areas, as well as multiple resource use areas' (SADC 1999b, 4). The objective of establishing TFCAs is clearly stated in Article 4(f) of the Protocol, in which the aim is '[to] promote the conservation of shared wildlife resources through the establishment of transfrontier conservation areas' (SADC 1999b, 6). This is also emphasised in Article 7 Sub-Section 5a, which states that 'State Parties shall, as appropriate, establish programmes and enter into agreements to promote the co-operative management of shared wildlife resources and wildlife habitats across international borders' (SADC 1999b, 12). The Protocol has since been used as the reason for member states to sign MoUs and treaties for the establishment of TFCAs.

The BSP commissioned a study on the development of transboundary natural resource management (TBNRM) areas in southern Africa, which was published in 1999 (Griffin *et al.*, 1999). Of significance to our discussion is that the study identified 17 potential terrestrial TBNRM areas, namely, Iona/Skeleton Coast, Okavango/Caprivi/Chobe/Hwange, Kgalagadi, Mana/Zambezi/Cahora Bassa, Kruger/Zinave/Banhine/Gonarezhou, Niassa/Selous, Ais/Ais Richtersveld, Drakensberg/Maloti, Malolotja, Maputo/Mlawula, Ndumu/Tembe/Maputo, Tuli Bloc, Chimanimani, Nyika/Nyika, Kasungu/Lukusuzi, Lengwe, and Mwabvi (Griffin *et al.* 1999). These are the same areas that the PPF had already proposed or showed interest in. In fact, the team leader of the study, John Griffin, acknowledged that the PPF supported the study with databases. It could be defensibly argued that the TBNRM areas proposed by Griffin *et al.* (1999) reaffirmed the peace parks map of the PPF. In that way, the PPF, though operating outside the formal structures of SADC, was able to promote its peace parks sites in the region through SABSP, which had the blessing of SADC structures.

Effectively, the PPF operated through DEAT and SANParks in South Africa. It entered into a trilateral agreement with the two institutions in 2002. In that agreement, DEAT's roles and responsibilities included: the proclamation and designation of areas forming part of the TFCAs in South Africa; co-ordination and facilitation of co-operation between the three tiers of government structures, parastatals and other agencies; management of Cabinet process; formalisation and implementation of policies; and being the lead agent in negotiations and entering into international agreements. For its part, SANParks' roles included: participation in conclusion of international agreements; management of co-governance arrangements; and formalising empowerment deals and ensuring community participation. In addition to fundraising for TFCAs, the PPF was, and still is, responsible for: the promotion of the development of TFCAs on a commercial basis; international marketing; and funding the salaries of TFCA Project Co-ordinators. The trilateral agreement empowered DEAT to play a pivotal role in the formalisation of TFCAs. Unsurprisingly, the former Minister of DEAT, Valli Moosa, used the departmental platform to champion the development of TFCAs in southern Africa.

I reiterate that though SADC structures, particularly its Sectoral Office in Malawi, and SADC Protocol on Wildlife Conservation and Law Enforcement recognised the establishment of TFCAs as strategic for nature conservation in the region, SADC did not initiate TFCAs in the region. Evidently, there have been delays in formally adopting existing TFCAs as the SADC programme of action. At the time of writing (i.e. May 2006) there was a proposal for establishing the Southern Africa Sub-Regional Environmental Action Plan (SREAP) as a regional expression of NEPAD's Environment Action Plan. As we have noted in Chapter 4, NEPAD has already endorsed the establishment of TFCAs as part of its continent-wide response to challenges posed by the threats to biodiversity and by the high level of poverty. As the Draft Document on SREAP (2006, 34) states:

> the purpose of implementing the action plan shall be to enhance the human, institutional, organisational and technical capacities of SADC sub-regional economic grouping and its partners for effective implementation of NEPAD environmental action plan and improve the ability to mainstream environmental considerations into other economic development streams of NEPAD.

Against this backdrop, TFCAs will be promoted as a continental rather than a regional initiative. The implications of linking TFCAs with NEPAD are threefold. Firstly, regional blocs and individual African states are most likely to adopt TFCAs as a way of implementing NEPAD. In the absence of the NEPAD framework, TFCAs are easily seen as a South African initiative that has been pushed through SADC, because of South Africa's dominant economic and political position in the SADC region and the continent as a whole. Linking NEPAD with TFCAs removes the 'South African factor' from the TFCAs, thereby promoting TFCAs as a continental rather than a South Africa-driven initiative. The second implication is that donors will be able to persuade regions and individual countries to establish TFCAs as a NEPAD initiative rather than a foreign imposition with the usual suspected conditionalities. Thirdly, as part of NEPAD, the failure of TFCAs to deliver on the promises of protecting biodiversity and socio-economic development would be laid squarely on African governments, all the more because NEPAD is touted as a home-grown programme that is owned and directed by enlightened African leaders.

This chapter has highlighted the regional contexts that have impacted on the establishment of TFCAs in Africa. Regionalism in Africa emerged as a differentiated process, which sought to promote socio-economic development in a post-independence political environment. As elsewhere, the main outcome of regionalism in Africa was the formation of regional blocs, which operated under different social, economic and political circumstances. With regard to the establishment of TFCAs, southern Africa stands out as the region leading the establishment of TFCAs in Africa, raising the question of the part played by regionalism in the development of TFCAs in that region. As I have suggested throughout this chapter, southern Africa is not necessarily unique in terms of the ecological and political reasons for TFCAs. Ecoregions are found in different parts of Africa, and African governments have co-operated over transboundary natural resources in one way or another. Politically, the ideology of pan-Africanism and the concerns with the divisive colonially inscribed boundaries permeated the entire continent. All these conditions have been used to promote TFCAs, yet southern Africa is the only region in which states are preoccupied with the establishment of TFCAs. This is so, mainly because southern African states embrace two different forms of co-operation, one non-spatial, the other 'space-producing'. The first one refers to co-operation on issues such as trade that do not produce fixed spatial entities. The second one, space-producing, refers to, or even demands, the establishment of a geographic area over which control is exercised as in the case of TFCAs. The PPF exploited SADC environment and structures to promote the establishment of TFCAs in the southern African region.

Endnotes

1 Unification through regionalism was not the only option available in early post-independence Africa. For example, Kwame Nkrumah of Ghana promoted the view that African unity should be achieved through the formation of the United States of Africa rather than regionalism.

2 The Convention is underpinned by the Governing Council decision 14/26 of the United Nations Environment Programme. 'The principal objectives of the Convention … are the conservation and sustainable use of biological diversity, and the fair and equitable sharing of benefits arising from its utilisation' (Secretariat of the Convention on Biological Diversity 2001, xvii).

3 Notably, the Conference for the Adoption of the Agreed Text of the Convention on Biological Diversity was held in Nairobi (Kenya) on 22 May 1992 (Secretariat of the Convention on Biological Diversity 2001).

4 The UNEP (2002) considers the northern Africa sub-region as made up of Algeria, Egypt, Libyan Arab Jamahiriya, Morocco, the Sudan and Tunisia.

5 www.iucn.org/themes/wcpa/region (accessed 08/3/2006).

6 In the work of the IUCN, North Africa and the Middle East are treated as one region.

7 Ranked by extinction risk score.

8 Discussion interview, 5/4/2006, Dar es Salaam.

9 I am thankful to Christine Noe for bringing this to my attention.

10 The declaration of an intent to form the Federation was signed by Jomo Kenyata (Kenya), Julius Nyerere (Tanzania) and Milton Obote (Uganda) in Nairobi on 5 June 1963 (Nkonoki 1983).

11 The member states of IGADD are Djibouti, Eritrea, Ethiopia, Kenya, Somalia, Sudan and Uganda.

12 The IGADD Heads of State and Government endorsed the decision to expand IGADD's area of operation on 18 April 1995 (www.igadregion.org).

13 The other two priority areas were: (i) conflict prevention, management and resolution and humanitarian affairs; and (ii) infrastructure development.

14 The EAC was revived in the 1990s and the Treaty for the Establishment of the East African Community was signed on 30 November 1999 in Arusha, Tanzania.

15 These are: banning all authorised cutting of native montane; maintaining the wildlife corridor; monitoring the discharge of the major rivers and streams; expanding Kilimanjaro National Park to include all of the montane forest; developing an integrated conservation plan that should define priority land uses and activities for Mount Kilimanjaro; developing extensive afforestation activities; designing long-term research projects that address resource problems; and planting several corridors of native vegetation (Newmark 1991).

16 The IUCN generally supports the development of TFCAs.

17 The time frame for the study was December 2001.

18 According to Baldus *et al.* (2003) the reserve represents more than 5% of Tanzania and is the largest uninhabited protected area in Africa.

19 It was formed as a coalition of the African Wildlife Foundation, Fauna and Flora International and WWF (Rainer *et al.* 2003).

20 Ministers Protais Mitari of the Ministry of Commerce, Industries, Investments Promotion, Tourism and Co-operatives (Rwanda) and Jovino Akaki Nyumu of the Ministry of Tourism, Trade and Industry (Uganda). The DRC was the only country represented by the Ministry of Environment, Nature Conservation, Water and Forests, under Minister Anselme Enerunga.

21 Benin, Burkina Faso, Cape Verde, Côte d'Ivoire, The Gambia, Ghana, Guinea, Guinea Bissau, Liberia, Mali, Mauritania, Niger, Nigeria, Senegal, Sierra Leone and Togo.

22 www.wrcu.ecowas.int (accessed 10/3/2006).

23 Reached at the Special Event Meeting of the Sahel and West African Club that was held in Accra on 20–21 May 2002.

24 In fact, states in the region were divided along the capitalist West and the communist East camps.

25 Angola, Botswana, Lesotho, Malawi, Mozambique, Swaziland, Tanzania, Zambia and Zimbabwe.

26 Known as the Lusaka Programme of Action.

27 Member states signed a Memorandum of Understanding that codified all SADCC decisions relating to the institutional arrangements.

28 In the following year (i.e. 1985) the ANC (Oliver Tambo), PAC (Johnson Mlambo) and SWAPO (Sam Nujoma) attended the Summit in Arusha.

29 For example, Botswana, Lesotho and Swaziland were members of the Southern African Customs Union, which was led by South Africa, while these countries also belonged to SADCC.

30 This deadline was not met.

31 The remaining four objectives were to, evolve common political values, systems and institutions; promote self-sustaining development on the basis of collective self-reliance, and the interdependence of member states; complementarity between national and regional strategies and programmes; and promote and maximise productive employment and utilisation of resources of the region (SADC 1992).

32 According to Chapter Seven, Article 21.

33 The year in which South Africa was welcomed as the 11th Member of SADC.

34 Lesotho was the first SADC member state to ratify the protocol on 30 July 1996.

35 The term TFCA was not even used in the Protocol. The concern is with the drainage basin, which the Protocol defines as 'a geographical area determined by the watershed limits of a system of waters including underground waters flowing into a common terminus' (SADC 1995a, 2).

36 The functions of the River Basin Management Institutions with regard to the protection of the environment are to: promote measures for the protection of the environment and the prevention of all forms of environmental degradation arising from the utilisation of the resources of shared watercourse systems; assist in the establishment of a list of substances whose introduction into the waters of a shared watercourse system is to be banned or controlled; promote environmental impact assessments of development projects within the shared watercourse systems; and monitor the effects on the environment and on water quality arising from navigational activities (SADC 1995a).

[37] There is a weak link between the protection of water resources and the idea of a national park.

[38] Discussion interview, 5/4/2006, Dar es Salaam.

[39] Personal communication with Muleso Kharika, 8/3/2006, Pretoria.

[40] Discussion interview with Professor Paul Maro, 5/4/2006, Dar es Salaam.

[41] Personal communication with Muleso Kharika, 8/3/2006.

[42] The idea of a focal point came as a result of the need for co-ordination among sectors dealing with biodiversity.

[43] SADC Ministers of Food, Agriculture and Natural Resources approved the functions of the Biodiversity Focal Point in Mauritius in June 1997 (SADC 1997).

[44] Including inland and marine fisheries, forestry, agricultural research, livestock production and land management.

[45] Botswana, Lesotho, Namibia, Swaziland and Zimbabwe.

[46] SANF carried out wildlife surveys in Malawi and also contributed to the development and management structures of national parks in that country.

Chapter 6

Silencing community struggles

Southern Africa as a region on which the experiment of TFCAs has been carried out is characterised by a long history of protests over loss of ownership and access to natural resources, especially land (Mbeki 1984; Ranger 1999). Such protests were visible at different historical moments. Clearly, the struggles for land rights were waged against colonists[1] and have continued in different forms into the post-independence era.[2] Protected areas such as national parks have not been immune to those struggles, hence they have been, and continue to be, zones of conflict between state agents and local residents (Kepe *et al.* 2005). Despite the history of community struggles for natural resource rights in southern Africa, there has not been any visible protest against the establishment of TFCAs in the region. This is surprising because TFCAs involve some of the national parks that have been contested by local communities and some of them also directly affect the surrounding communities. Their disenfranchising tendency, to borrow from Dzingirai (2004), does not seem to arouse protests among the disenfranchised. This begs the question of why communities living inside and around areas designated for TFCAs have not protested against the creation of those TFCAs.

The purpose of this chapter is to account for the lack of community protests against the TFCAs. It seeks to offer a theoretical explanation that is grounded on experiences and practices in TFCA sites. Theoretically, the chapter explains the lack of community protests by drawing on insights from a well-established body of work on scale analysis, the aim being to locate the community within the construction of the scale at which TFCAs are established. In other words, we need to observe processes that underscore the creation of TFCAs at the appropriate level so as to understand how those processes open or close the spaces of engagement by communities.

The chapter refers to constructions of scales to make three related arguments. Firstly, it argues that TFCAs manifest particular constructions of scales that disengage local communities, hence those communities are unable to contest or shape TFCAs. That is to say that the local communities are marginal to the processes that produce TFCAs. Secondly, it argues that TFCAs involve the double rearticulation of political scales. Whereas the need for the joint protection and management of biodiversity and fund raising forces the state to scale up its regulatory functions to a regional entity, the local demands and struggles for natural resources result in the scaling down of the functions of the state. Individual states remain responsible to their citizens in terms of their constitutions. Apart from signing the necessary legislative documents, states play a crucial role in realigning their policies to the ideals of TFCAs. The third argument is that at the sub-national level, community-based natural resources management projects mediate conflict over natural resources while, at the same time, ensuring that communal land functions as corridors for TFCAs. By their very nature, TFCAs require the participation of states and other landholders and resource users. Some comments on the constructions of scale are helpful in contextualising the arguments I make in this chapter.

Constructions of scale: implications for environmental issues

There seems to be consensus on the view that scale is not ontologically pre-given but is socially constructed (Brenner 2001; Marston and Smith 2001). Nevertheless, the debate on how different actors produce scales and the implications of the arrangement of scales on different processes is inconclusive. Basically, there is no agreement among scholars about how scales can best be arranged to understand the important roles that they (i.e. scales) play in different processes (Beauregard 1995; Thrift 1999; North 2005). The disagreement is indicative of different views on how scales are produced and work in different settings. Of significance to our discussion is how questions of scales have been related to environmental matters. Brown and Purcell (2005), McCarthy (2005) and Sayre (2005) lament that the insights gained from the debate on scale have not been successfully integrated into political ecology. Sayre (2005) is worried that there has not been comparison of issues of scale in ecology and geography, and suggests that each discipline requires insights from the other. The possibilities of integrating scalar debates between ecology and human geography lie in the fact that geography's epistemological insight into the production and significance of scale could be applied to ecology, while ecologists can help resolve 'the scale question' in critical human geography (Sayre 2005). In emphasising the importance of scale in ecology, Sayre (2005, 277) referred to Lee's view that, 'when human responsibility does not match the spatial, temporal, or functional scale of natural phenomena, unsustainable use of resources is likely, and it will persist until the mismatch of scales is cured'.

Brown and Purcell (2005) are of the view that research in political ecology would benefit from more explicit and careful attention to the question of scale and scalar politics. Proceeding from the familiar view that scale is socially produced through contingent political struggles, they sought to relate discussions about scale with ideas of local community participation and benefits sharing. They are particularly concerned that there is a strong belief among political ecologists that local scale arrangements are inherently more likely to have a desired effect on local communities than activities organised at other scales.[3] In other words, political ecologists suffer from a 'scalar trap', which privileges the local scale. In rereading the Brazilian Amazon through the lens of scales, Brown and Purcell (2005) argued that scalar arrangements were fixed, unfixed and refixed in a process of continual political struggle in order to achieve specific agendas. In their view, the local producers in the Brazilian Amazon were able to jump from the local to the international scale by forming alliances with international capital interests (e.g. the World Bank) and international NGOs such as WWF. This jumping of scale, which they term glocalisation, has led to the construction of a new scale that is more beneficial to the local producers than the local scale.

McCarthy (2005, 734) also supports the relevance of glocalisation in understanding the politics of scale, arguing that 'environmentalists have vigorously and successfully promulgated both positive and normative claims about the scales at which environmental problems should be conceptualised and environmental governance should occur'. McCarthy makes two important observations that are relevant to the discussion in this chapter. The first observation is that the current theorisations of scale in radical geography have paid little attention to environmental NGOs and movements as important actors in scalar politics. This observation challenges the tendency to regard the state as having the central and unique role in the construction of scales. The second observation is that indigenous groups and international environmental NGOs have developed scale-jumping strategic alliances to construct new spaces of engagements. In support of this, McCarthy (2005) referred to the Methanex case in California,[4] in which environmental NGOs made submission to the tribunal, a move which resulted in the restructuring of the scope and number of participants in that case. Under NAFTA, the Canadian-based Methanex Corporation can initiate disputes against foreign states directly without the approval of the Canadian government, a situation that could disable Canada's participation in the resolution of the dispute. McCarthy's point is that agreements with the capacity to supersede domestic regulations are a clear instance of the rescaling by capital to escape domestic constraints. In the Methanex case, environmental NGOs disrupted that rescaling process to allow voices that would otherwise not be heard.

The examples of the Brazilian Amazon and Methanex referred to above suggest that local residents participate in the construction of scales by forming alliances with international environmental NGOs. These examples are correct as far they go, but leave out another important role played by international environmental NGOs, which is to marginalise local residents. Chapin (2004) has accused the three main international environmental NGOs, namely, Conservation International, The Nature Conservancy and WWF, of ignoring the very indigenous people whose land they claim to protect and conserve. Brown and Purcell's (2005) and McCarthy's (2005) work seems to suggest that local residents assert their power through glocalisation as if their engagements at other scales are not beneficial. Local residents could equally engage with national and sub-national structures of power in their attempts to extract the same gains that accrue through glocalisation. As I will be showing below, TFCAs create an environment in which local residents are exposed to forces operating at and through various scales, leading to a form of glocalisation that weakens the residents' position.

TFCAs as a scale of marginality

In contrast to the hierarchical ordering of scales, which begins mainly from the local, the starting point for TFCAs is the supranational scale. I refer to this scale in this chapter as a regional scale. The regional scale for TFCAs is embedded in the idea of crossing international boundaries, and is supported by the metaphor of peace parks between and among nations. That is to say that the actors involved in the constructions of the scale for TFCAs share common scalar interests, though these are pursued from different angles (see Chapters 2, 3 and 4). I argue that local residents have had little or no contribution to the construction of the regional scale for TFCAs.

The double rearticulation of political scales

A growing body of work has been critical of a state-centric approach that privileges the state as the main player in the construction of scale (McCarthy 2005). In the case of TFCAs, the state plays a critical role in connecting both regional and global actors, and local residents to the goal of transfrontier conservation. Nevertheless, the ways in which these actors are connected are characterised by imbalances. Both weak and strong states agree to the need for the joint protection and management of biodiversity in order to raise funds and to develop appropriate structures of management and control. In this sense the state scales up its regulatory functions to a regional entity. In southern Africa, the process of scaling up the functions of the state are facilitated in two main ways, the first being that donors are increasingly favouring regional rather than national development projects, as is the case with funding from the PPF,[5] GEF and USAID.[6] Southern African states organise themselves under the banner of SADC in order to access those funds (see Sidaway 2002). The involvement of SADC leads to the second route in the scaling process, namely, the development of protocols that bind states to a common vision while, at the same time, committing governments to developing national policies that articulate regional goals and ideals. Of significance to our discussion here are the Protocols on Wildlife Protection and on Tourism, which promote TFCAs. By signing these protocols, the governments of southern African states agree that their national policies should be redesigned in line with regional interests and goals. This is precisely the point at which there is competition between national and regional interests. With regard to TFCAs, governments have tended to support regional interests at the expense of their citizens, particularly those living inside or adjacent to areas designated for TFCAs. Mozambique and Zimbabwe provide a good example of how this happens.

The Great Limpopo and local communities in Mozambique

The official process of creating the GLTP that took place between 1990 and 2002 occurred at the national and supranational levels, with funding from GEF, GTZ, governments and the private sector. For Mozambique to participate fully in that transfrontier park, it had to elevate the status of the Coutada 16 from a hunting ground to a national park, a process that culminated in the creation of Parque Nacional do Limpopo (hereafter called Limpopo National Park) in 2001. It could be argued that Limpopo National Park was part and parcel of the conditionality[7] imposed on Mozambique and, as in such conditionalities, the state tends to ignore local interests in favour of international ones.[8] The World Bank, through its Global Environmental Facility, identified Coutada 16 as one of the areas that, 'when coupled with the rapidly growing potential for tourism in the region, represents unique opportunities to combine biodiversity conservation and sustainable development' (World Bank 1996, 3). Leaving local interests aside for the moment, the Mozambican government embarked on legislative and institutional changes that, from the mid-1990s onwards, were supportive of the creation of TFCAs, including the GLTP. For example, the government introduced the National Environmental Management Programme in 1995 and subsequently created the Ministry for Co-ordination of Environmental Affairs. Conceptually, these changes were introduced to realign the activities and functions of the Mozambican government to global protocols such as the CBD, which that government had signed in 1995. Against this backdrop, the creators of the GLTP predominantly operated at, and between, the national, regional and global scales.

Global bodies, such as the IUCN, that participated in the creation of the GLTP failed to assist local residents to jump the scale. That is to say that the normal process of glocalisation was interrupted. A number of reasons can be advanced to explain this. For example, the Regional Office of the IUCN (IUCN-ROSA) based in Zimbabwe was ambiguous towards the creation of TFCAs[9] in the region and was preoccupied with CBNRM projects when the idea of the TFCA was being nurtured. It has been suggested that the promoters of TFCAs attempted to change the focus of the IUCN-ROSA from community-based projects in Mozambique to that of TFCAs, thereby creating confusion over the purposes of CBNRM projects in that country.[10] Similarly, other NGOs that participated in nature conservation projects in the GLTP site in that country, through the ticket of community empowerment and development, were also entangled in the GLTP, which, as I will show, was marginalising the community. Confusion over the roles that the CBNRM projects were to play in the creation of the GLTP was reflected among the donors. For example, the PPF, IUCN, USAID, GEF and the German Development Bank funded the creation of the Limpopo National Park, but both the World

Bank and German Development Bank were later reluctant to continue funding that park, because of ways in which the affected communities in Mozambique had been marginalised from the GLTP process. In 2004, the World Bank commented that, 'despite their efforts to resolve the fate of the communities living in Coutada 16, and the assurances given by [the Government of Mozambique] that the issue would be adequately resolved prior to any action, in 2001, the [Limpopo National Park] was gazetted as a national park and some animals were allowed to enter the park area. This created a lot of media attention and controversy around the project' (cited in Spenceley forthcoming, 168). This begs the question of how the World Bank funded the GLTP without full knowledge of the impact that the GLTP initiative would have on local populations. Arguably, the World Bank was aware that there were people living inside the TFCA that was later to become the GLTP. Its grant of US$5 million towards the pilot project in TFCAs in Mozambique was meant to 'contribute to poverty reduction by assisting local communities *inside* and around the conservation areas' (World Bank 1996, n.p.).

It follows that local residents in Mozambique were left behind in the process of creating the GLTP. In its survey of the community's knowledge and awareness of the GLTP in 2002, the University of the Witwatersrand's Refugee Research Programme concluded that communities in Mozambique had little or no knowledge of the GLTP (Wits Refugee Research Programme 2002). Despite these findings, the pro-GLTP consultants, Suni/CREATE, promoted the view that residents knew that their areas were within the boundaries of the Limpopo National Park – a park which forms part of the GLTP. There is a general consensus among analysts of the GLTP that communities were left outside the process of creating the GLTP for a long time despite public pronouncements on the GLTP as a pro-community initiative. It has been suggested that some high-ranking government officials opposed the idea of community participation in the GLTP in the boardrooms, but delivered pro-community speeches on the public platforms, as a marketing excise.[11] In addition, there has not been a common approach towards local communities affected by the creation of the GLTP. This is evidence of the lack of support for local communities at national and regional levels.

In practice, the Mozambican government has to carry the burden of addressing the consequences of community marginalisation in the Limpopo National Park. This involves resettling approximately 6500 people from the Limpopo National Park. According to the first manager of Limpopo National Park, Gilberto Vincente,[12] the position of the Mozambican government is that people will be resettled outside the Park on a voluntary basis, possibly to an area south of Massingir Dam, but within the Gaza District (Figure 6.1). He commented that:

> there will be no forced removals, à la the apartheid government's efforts in the early days of Kruger [National Park] … This time we need to get it right … If we don't, they [people to be removed] will return in no time at all and try to reclaim their land … On a positive note, however, most villagers are in agreement with our plans and the only remaining hurdle is finding alternative land which is acceptable to all parties (Vincente, cited in Edwart-Smith 2005, 27).

Apparently, the reason for relocating people *en masse* to the same place is to keep them under the control of their traditional leaders. There is fear among traditional leaders that resettlement could split their subjects, thereby diminishing their power base. Traditional and elected leaders in the area are divided on the issue of resettlement. Arguably, the pro-government elected leaders in the area are likely to toe the government line and promote resettlement as the solution to the problem of parks versus people.[13]

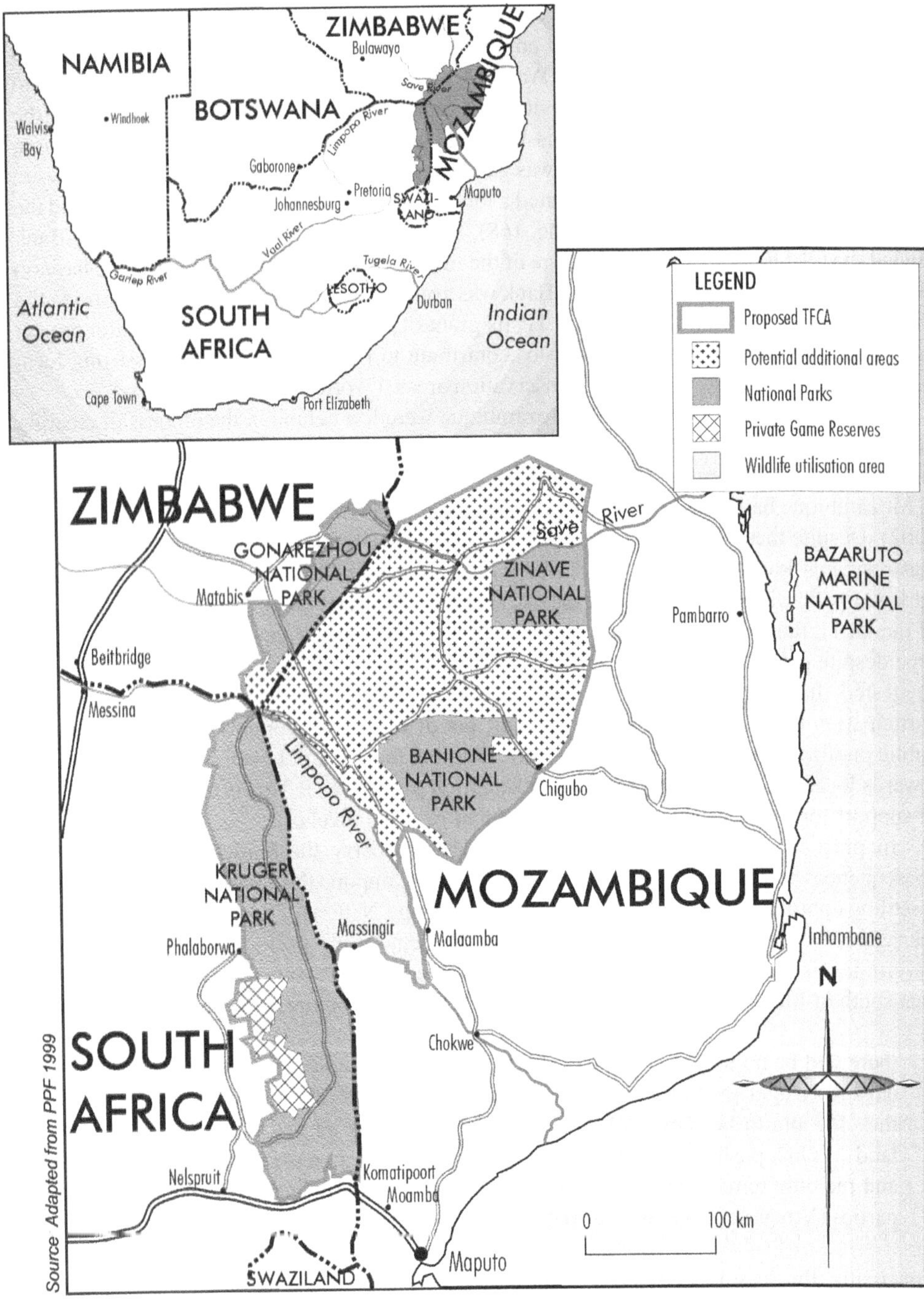

Figure 6.1 The GLTP and environs.

The need to relocate people is presented by officials and environmentalists as an accidental outcome of the process of creating the GLTP. Vincente reasoned that when the idea of the GLTP was conceived, it was anticipated that the existing national parks would be joined together. However, it was difficult to create corridors between national parks that were far apart and that have human settlements between them, as was the case between national parks in South Africa and Mozambique. Consequently, Coutada 16 was upgraded from a hunting concession to IUCN Category II national park. In the light of the discussion above, the 'accidental discovery' of people in the park is a fallacy, because all the geographical designs for the GLTP incorporated the areas that had been occupied by people. But, the creators of the GLTP were silent about the fate of people living inside the GLTP in order to avoid scaring off donors. The silence was also meant to weaken any possible resistance from the local communities and international pressure groups at a crucial time when creating the GLTP was proving to be a difficult undertaking at the intergovernmental level.

A number of resettlement options have been suggested for communities on the Mozambican side of the GLTP. These include providing alternative serviced sites on which people would be resettled, fencing people into enclaves in the park or allowing people and wildlife to live together in the park. The Mozambican government endorsed the resettlement option in 2004, providing incentives that would make the option appear voluntary. Ferreira (forthcoming) has indicated that the German Development Bank (KfW) will contribute 6 million Euros towards resettlement, and that a draft proposal for the resettlement of several small agricultural villages was on the table for discussion in mid-2005. Donors who were becoming reluctant to fund the GLTP, because of the controversy surrounding resettlement, seem to have found another avenue for participating in the GLTP through community-based projects. Arguably, the financial assistance from the German Development Bank towards resettlement is meant to present the Bank as pro-community, while ensuring that the GLTP develops as planned. Similarly, the Swiss-based NGO, Helvetas, and USAID have jointly funded the Covane Community Lodge, which was opened in May 2004. It is assumed that communities will use the lodge to benefit from ecotourism. Projects of this nature divert the attention of the public and the local communities away from the issue of resettlement to that of community development, thereby ensuring that the design and ambitions of the GLTP remain intact.

The Great Limpopo and local communities in Zimbabwe

Drawing from his research on the Sengwe community in Zimbabwe, Dzingirai (2004) concluded that TFCAs, such as the GLTP, represent a process of disenfranchisement at large. By this he meant that TFCAs do not only lead to the loss of community control over, and access to, natural resources,[14] but also result in large-scale resettlement of communities. As in the case of Mozambique, the displacement of people in Zimbabwe 'is one that does not happen in a rhetorical vacuum. [But] is sanitised in ways that presents it as less disruptive' (Dzingirai 2004, 8). How this happens on the ground will become clear below.

The participation of Zimbabwe in the GLTP depended mainly on linking Gonarezhou National Park (Zimbabwe) to the Kruger National Park (South Africa). Unlike in situations where a TFCA is established by simply joining contiguous national parks,[15] Gonarezhou could only be joined to the Kruger National Park by means of a strip of land occupied by different communities in Zimbabwe. There is no common boundary between Gonarezhou National Park and either the Kruger National Park or the newly formed Limpopo National Park (Mozambique). According to the Italian-based NGO, CESVI, Zimbabwe had to use the

'Sengwe Corridor' and the 'Limpopo strip' (Figure 6.2) to commit the Gonarezhou National Park to the GLTP and to comply with the international treaty of GLTP (CESVI 2005). Indeed, the Treaty of the Great Limpopo Transfrontier Park (2002) defined the areas of the GLTP in Zimbabwe as: Gonarezhou National Park; Malipati Safari Area; Manjinji Pan Sanctuary; and the community areas, which constitute the biodiversity corridor linking Gonarezhou to the Kruger National Park (Figure 6.2).

Obviously, the Treaty did not spell out how that corridor would be created, thereby leaving the responsibility of establishing the corridor to the Zimbabwean government. In terms of the theme of this chapter, by signing the Treaty, the Zimbabwean government was bound to create the corridor, thereby promoting interests at the supranational level. Practically, there was no way in which the government could have been supportive of community opposition to

Figure 6.2 TFCAs in south and south-eastern Zimbabwe.

TFCAs. The same applies to the NGOs that have been involved in some kind of nature-based community development projects in the envisaged corridor. For example, the Parks and Wildlife Management Authority, CESVI and African Wildlife Foundation have been partners in the creation of the corridor. They jointly organised a Strategic Planning Workshop for the Sengwe Corridor in September 2002, the aim being to develop strategies for creating the corridor. Their common desire is to physically link Gonarezhou and Kruger National parks in order to facilitate wildlife movement, tourist flows and regional economic development. The policies in Zimbabwe that are key to this process are Wildlife Policy, Wildlife Based Land Reform and the Policy for Tourism Development. The Wildlife Policy is relevant to the GLTP as it specifically refers to the wildlife management and utilisation, and forms the basis for developing wildlife-based land-use systems. However, it does not provide options for creating the Sengwe–Tchipise Wilderness Corridor. The Corridor could be established either by contractual parks or co-management arrangements, options that are not provided for in the Wildlife Policy. In the absence of these legal instruments, different options for creating the corridors have been pursued. For example, the Transfrontier Park Technical Committee proposed a strip 15–26 km wide, which implied the resettlement of approximately 15,000 people (CESVI 2005). The Sengwe community is opposed to the resettlement. According to Dzingirai (2004) neither external nor internal relocation is practical or desirable. External relocation implies that the Sengwe community will have to be relocated to commercial farms or some unused land in the nearby villages. Given the political sensitivities of land reform in Zimbabwe, the relocation of the Sengwe community to commercial farms for the sake of the corridor will probably bring the GLTP into disrepute. Equally, internal relocation is not feasible, because people have been migrating out of Sengwe proper to areas adjacent to protected areas in search of livelihoods (Dzingirai 2004).[16]

This means that the creation of the corridor requires strategies for accommodating local communities as part of the corridor. CESVI (2005) has suggested four planning scenarios for the creation of the corridor and their implications for the GLTP. The first one is to maintain the status quo at the peril of the local economy and biodiversity. In other words, the social, economic and environmental consequences of this scenario are all negative. The second involves developing the tourism industry with little, if any, integration between land tenure systems. This situation would, according to CESVI (2005), lead to local communities and resettled farmers reverting to an agro-livestock based land-use system. It would also promote low-key non-consumptive tourism while, at the same time, limit options for the development of Gonarezhou National Park. The result would be a fragmented tourism development. The third scenario is tourism development, which focuses on Limpopo/Sengwe/Mabalauta and Gonarezhou National Park.[17] The last scenario, which is considered the most robust of all options, involves the launching of the Integrated Tourism Development for the south-eastern lowveld. To achieve this goal, the Parks and Wildlife Management Authority and the Sengwe–Tchipise communities will enter into an arrangement to jointly manage the development zone. Similar arrangements will be made with other affected communities,[18] conservancies and the private sector. CESVI (2005) suggests that, while the linkage between Gonarezhou National Park and its counterparts in South Africa and Mozambique is a necessity, the process of creating the corridor should be broadened to accommodate current and potential interests in developing the south-eastern part of Zimbabwe. In 2005 plans were already underway to join the Sengwe Corridor and 'Limpopo Strip' into the Sengwe–Tchipise Wilderness Corridor (Figure 6.2), to be gazetted under Zimbabwe's Town Regional and Country Planning Act.

Arguably, the proposal for the wider wilderness corridor goes beyond enabling the government of Zimbabwe to participate in the GLTP, but is meant to resolve similar problems that would arise from the establishment of the Limpopo/Shashe TFCA at the Botswana–South Africa–Zimbabwe border. The creation of that TFCA will affect the Machuchuta, Maramani and Masera communities (see Figure 6.2), whose communal land constitutes more than half the total area of the TFCA component in Zimbabwe. The population of the three communal lands has been estimated at 10,500 (CESVI 2003). There is silence on how these people will be accommodated in the TFCA, partly because the Zimbabwean government has delayed the signing of the Memorandum of Understanding. Nevertheless, the proposal for the Sengwe–Tchipise Wilderness Corridor pre-empts the linkage between the Limpopo/Shashe and developments in the Sengwe Corridor. Conceptually, the Sengwe–Tchipise Wilderness Corridor offers a sub-regional solution to problems that affect specific communities. That is to say that in the absence of solutions at the national and international levels, a sub-regional process is being promoted to diffuse a potential wave of local protests. The process 'de-localises' potential protests by encouraging an ecotourism dialogue among communities that would be affected by TFCAs in southern and south-eastern Zimbabwe. In this way, communities engage with the TFCA process from an angle that has been defined for them. That is, the involvement of communities, as a sub-regional collective, in the development of TFCAs did not originate from the mobilisation by the communities themselves. Instead, it emerged from a process that sought to direct communities to the goals of TFCAs. In this sense, communities participate in, and for, their own marginalisation.

The example of community marginalisation in the GLTP, particularly on the Mozambican side, has attracted the attention of analysts of TFCAs in southern Africa. The reason for this is obvious: the affected communities are located within the most publicised TFCAs in Africa. Nonetheless, the marginalisation of communities in the GLTP is not unique to a particular TFCA in southern Africa. That process is built into the conception, design and practices of TFCAs in the region. Most, if not all, TFCA maps in the region took the form of 'classified documents' in their initial stages. They were not made public and communities did not know that their areas would be incorporated into TFCAs. In official circles, there were those who feared that publicising TFCA maps during the conception stage would agitate people.[19] The point here is that TFCA maps are imposed on local communities when the deal is done between governments and donors, diminishing the chance of communities influencing the TFCA processes.

Theoretically, the local scale is peripheral to the TFCA process, rendering the common notion of glocalisation in which local communities forge links with global players almost untenable. Stated differently, the process of jumping scales is disrupted by the removal of the scale (i.e. local) from which the jumping should normally start. This is not to argue that there is no glocalisation in transfrontier conservation. Rather it is to suggest that glocalisation does not have to be limited to the idea of jumping scales, because the scales are by and large connected and actors at different scales could push that process. The process of glocalisation takes different forms that are contingent on the issues and actors at hand. With regard to the TFCAs, the roles that are usually played by local communities in the glocalisation of environmental issues are taken over by elites, who are not necessarily members of those communities, but who are able to connect with actors above the local scale. The role that Anton Rupert played in promoting TFCAs demonstrates this point. He penetrated national governments and international institutions, while presumably promoting the interests of black local communities that he neither belonged to nor with whom he planned TFCAs.[20]

Hughes (2005) mooted the idea that glocalisation in TFCAs takes the form of scale-jumping by which southern African whites[21] hope to escape from black majority governments in post-apartheid southern Africa. His argument is that TFCAs 'allowed [these whites] to escape the confines of the nation and jump scales upward … Among these [whites], some have chosen – almost as an act of will – to identify with the continent as a whole, rather than with any one country or even with the southern subcontinent' (Hughes 2005, 160). In this sense the scale that is being jumped is the national one. While Hughes' attempt to bring the debate about the production of scale to the analysis of TFCAs is appreciated, his suggestion that the national scale is being jumped implies that southern African whites operate from the local scale.

In light of the discussion above, this constitutes a form of glocalisation, but Hughes neither clarifies the actors with which those whites are linked at the continental scale, nor explains why they would want to be 'stateless' citizens in the continent. Hughes' argument is also problematic, because the black elite has been participating in the establishment of TFCAs (Draper *et al.* 2004). Are those black people also escaping from the state and the sub-continent, and what would their reasons be? The suggestion that the national scale is being jumped is also questionable, because the processes by which TFCAs are produced do not eclipse actors at the national scale. In fact, TFCAs require the active involvement of actors, some of which operate at the national level. More crucially, to equate southern African whites with 'the local' is to ignore the presence of local black residents who are supposedly the main beneficiaries of the TFCAs. The local communities have a specific legal definition in TFCAs, which suggests a particular view of 'the local'. What appears to be happening in TFCAs is the reconstitution of the local so that local actors are not necessarily local communities as defined in the legal and official documents of TFCAs. For example, conservation agencies could be said to have taken the place of local communities in the creation of TFCAs. Conservation agencies in Botswana and South Africa championed the establishment of the Kgalagadi Transfrontier Park and were able to successfully link up with external and internal donors.

In summary, the kind of glocalisation taking place in TFCAs weakens the position of local residents who are central to the discourse of TFCAs while, at the same time, allowing the local elite, governments and NGOs to participate in the scaling processes. In addition, glocalisation in TFCAs works through the national scale instead of jumping it. For local communities, as in the cases of Mozambique and Zimbabwe, their demands and struggles for natural resources are handled at the national and sub-national levels rather than at supranational level where the TFCAs' process is most active. In this sense individual states are responsible for executing tasks at the national and sub-national levels that are supportive of a process that takes place at higher levels. This is clear in the development of state policies that are aligned to the ideals of transfrontier conservation. In the section that follows I show how state-driven nature conservation policies, which seemingly developed independently of the TFCAs, were variously realigned with the logic of an externally manufactured vision of the ecological landscape of southern Africa.

The congruence between state policies and the imagined geography of TFCAs

Globalisation might have challenged the modern geopolitical imaginations by rearranging the interconnectivity and functional boundaries of the world political map (O'Tuathail 1998), but imagined geographies continue to inform and shape contemporary political developments and processes. Thus, developments and processes occurring at various scales are continuously being enmeshed in an imagined future. That future is, as Visvanathan (1998, 239) put it, 'no longer fiction or fantasy. It is being colonised by an oracle of international civil servants who have mapped it with cybernetics and systems theory'. In practice, geographical imaginations have spurred and realigned activities in accordance with a projected future. A great deal of this has been learnt from imperialist imaginations that resulted in the growth of institutions and associations 'dedicated to propagandising the increasingly enfranchised masses into believing [that] colonial expansion was in everyone's interest' (O'Tuathail 1996, 21). Moreover, imperialist expansionism thrived on the imagined distant colonies, hence the colonisation of Africa required the constructions of images of the continent as an empty and refused land (Marks 1980; Mazrui 1980; Mudimbe 1988). Such an image (and there are many more) was not only used to justify European occupation and the imposition of Western values on the continent, but also provided the context in which colonial policies evolved. In other words, out of an imagined continental space arose policies that sought to reconfigure that space in accordance with the imperial vision of the world.

State policies and doctrines therefore express the contours of particular political imaginations, as the history of geopolitics illustrates. O'Tuathail (1996) noted that geopolitics and its rootedness in an imagined earth informed national security state doctrines. The post-Cold War foreign policy of the United States is rooted in projections of the future world order and the role and position of the United States in it. The challenge therefore is to bring to intelligibility the raw material of political imaginations on the one hand, and how those imaginations are linked to processes and developments on the ground, on the other hand. Simply put, how can we be sure that a particular development, process or policy fits into some grand plan? In this section I grapple with this question by looking at the foundation of one of the latest imagined geographies of the African continent (Hughes 2005), TFCAs, and their connectedness with state policies. The African Dream, as TFCAs have come to be called, is an imagination of a connected ecological landscape of the continent (i.e. Africa) by means of conservation areas. My intention is to demonstrate that state policies in the region are linked to the imagined geography of TFCAs in three main ways. Firstly, governments formulated policies that pre-empted the TFCAs' environment. Secondly, protagonists of TFCAs appropriated existing state policies to draw their TFCA maps. Thirdly, governments have been encouraged to develop policies that realign land-use options in order to facilitate the drawing of a preconceived map. All these were made possible through the use of multiple sources in which the imagined geographies of TFCAs are embedded.

Government policies and TFCAs: the case of South Africa

The drive for ecotourism inside and outside TFCA-designated areas has given impetus to the policy of privatisation in national parks. For instance, the policy of privatisation in South Africa's national parks closely followed the development of the GLTP. The first private game concession near the Kruger National Park, Ngala, was established in 1990, the same year in which Rupert openly sold the transfrontier park idea to governments in the region. According to South African Airlink (2003, 8), Ngala 'heralded the first three-way partnership between the private sector (CC Africa), the state [SANParks] and a non-governmental organisation (World Wide Fund) and paved the way for the establishment of [transfrontier parks]'. Furthermore, the signing of the Memorandum of Understanding for the establishment of the GLTP and the first phase of the programme of privatisation in South Africa's national parks all took place in November 2000 (South Africa 2000; SANParks 2004). The privatisation policy was implemented in time (i.e. 2000) for the private sector to operate in the GLTP on the South African side. Through its privatisation programme, Operation Prevail, private companies invested R122 million in six sites in the Kruger National Park in 2001 (Ramutsindela 2004b). Although the programme is not limited to the Kruger National Park, that park accounts for approximately 60% of the number of concession areas in the country's national parks.

The official explanations for privatisation in South Africa's national parks was that: the democratic government will prioritise social expenditure above nature conservation; national parks should be run on business lines; and that SANParks' priority should be on managing biodiversity. Be that as it may, the International Finance Corporation (a subsidiary of the World Bank) influenced the process of privatisation in South Africa's national parks through its technical input. The lesson here is that privatisation in national parks in the region cannot sufficiently be disconnected from the ambitious African dream. I emphasise that concessions in South Africa's national parks were granted in anticipation of the role that the private sector would play in TFCAs in the region.

It should be noted that privatisation in the Kruger National Park was accompanied by other private sector investments outside the Park. For example, a 22-month project was underway to build the Kruger Mpumalanga International Airport (KMIA),[22] which became operational on 21 October 2002, two months before the signing of the Treaty of the GLTP. KMIA is owned by Primkop Airport Management and is officially described as 'the international gateway in Mpumalanga for direct access to the world-renowned Kruger National Park and the world famous Sabi Sands private game lodges in the Greater Kruger' (KMIA undated, n.p.). In its first years of operation, the KMIA had direct international flights to Australia, Britain, China, Switzerland and the United States. Arguably, the KMIA was built in order to fly international tourists directly to the GLTP. Evidence of this is found in the KMIA brochure, which sells the GLTP as one of the main attractions in the vicinity of the airport. Furthermore, the aircraft of South African Airways that flies from Cape Town International Airport to KMIA had the following words on its side: 'Flying the African Dream',[23] a description that resonates with TFCAs as an 'African Dream'.

Conceptually, the existence of an imagined map could pave the way for developing appropriate policies that shape activities for the realisation of that map. In this sense, concessions are an example of proactive policies that support the development of TFCAs. In addition to this strategy, the imagined geography of TFCAs also relied on state policies that were already in operation. Thus, TFCA protagonists have appropriated existing state policies to enhance their vision for TFCAs, as the case of Wildlife Management Areas (hereafter WMAs) in Botswana illustrates. The WMAs in Botswana were championed by the state, and developed from policies that were not linked to the TFCA idea at the time. However, WMAs were later to be embraced as critical raw material for the imagined geography of TFCAs in southern Africa.

Appropriating state policies for TFCAs: the case of Botswana

Wildlife forms an important strand in the history of colonial and postcolonial Botswana. Consequently, it was given attention when ideas of land-use planning diffused into the newly independent Botswana in the late 1960s. It should be noted that, in its initial stages, land-use planning in Botswana focused on the construction of small dams for planned settlements. It was aimed at bringing about a redistribution of cattle and people (Botswana 1968). In this sense, land-use planning led to the implementation of the Tribal Grazing Land Policy (TGLP) in 1975, which was to determine the shape of the countryside in that country. According to the Botswana Government, 'the TGLP was meant to enhance proper management and use of land so as to avoid overgrazing. The basic plan was to relocate the large cattle owners so that pressure on communal areas would be reduced' (Botswana 1997, 333). Furthermore, the policy formalised three categories of land, namely, communal grazing areas, commercial farming areas and reserved areas. The reserved areas 'were further divided into two categories; those with only a few cattle and areas reserved for alternate uses such as wildlife, mining and cultivation' (Botswana 2001, 197).

Of relevance to the discussion of this chapter is the emergence of areas reserved for wildlife uses, which are designated as WMAs. It should be emphasised that WMAs originated from intensive land-use zoning in that country and that they were based on the principles of sustained utilisation and co-ordinated management of wildlife outside the parks and reserves. The framework for these was laid down by the Wildlife Conservation Policy, which received official endorsement in July 1986. Their importance does not only lie in the area they cover (i.e. 20% of Botswana in 2001), but in the purposes they are meant to serve. First, they serve as buffer zones between conflicting land-uses, particularly the competition for resource utilisation between cattle and wildlife. The idea of buffer zones has gained global recognition as a mechanism for softening boundaries around protected areas. In theory, buffer zones are mechanisms through which the boundaries between parks and people are blurred. The second function of WMAs in Botswana is to open up the space for diversified wildlife utilisation by local populations, 'forming a rural base for commercial and non-commercial activities to provide employment and income opportunities for rural dwellers' (Botswana 1985, 255). Clearly, this is a departure from narrow and conservative ideas of preservation and law enforcement. Nonetheless, the question of whether local populations do actually benefit from WMAs has attracted a great deal of attention (Twyman 1998; Rozemeijer and Van der Jagt 2000; Mbaiwa 2004), not least because the Wildlife and National Parks Act of 1992 'authorises the Department of Wildlife and National Parks to use wildlife resources in any given Wildlife Management Area' (Botswana 1997, 333). Thirdly, WMAs provide migratory corridors for wildlife. As I have shown in Chapter 3, TFCAs are underpinned by the idea of migratory corridors that had been disrupted by colonial boundaries. In the case of Botswana, WMAs created the space for migratory corridors within the state, correctly implying that TFCAs should not be confined to the international domain only.

Notwithstanding their origin as a state policy on land-use and zoning, WMAs fit into the plan for TFCAs by linking different land-use types along the axis of TFCAs (see Figure 6.3). In this regard, Botswana stands alone in the region as the country that has provided TFCA ideologues with ready-made 'points of connectivity' for establishing TFCAs across the country. When the Botswana government became actively involved in TFCAs (through the Kgagaladi Transfrontier Park), all that was left was to perfect a system that was already in place. Subsequently, it had to resolve the problem of human settlements that were outside WMAs but within the envisaged TFCA map. To this end, the removal of the Basarwa from the Central Kalahari Game Reserve has something to do with ensuring that a 'wilderness path' is established from the south to the north of the country. Different explanations have been offered for the removal of the Basarwa. The Government argues that the Basarwa are entitled to the country's benefits of development like any other citizen of that country, and that they should be resettled from their scattered settlements into compact settlements in order for government to provide them with 'infrastructure and services in a cost effective and efficient manner' (Botswana 1998, 20). According to the *Mail & Guardian* (1999), the

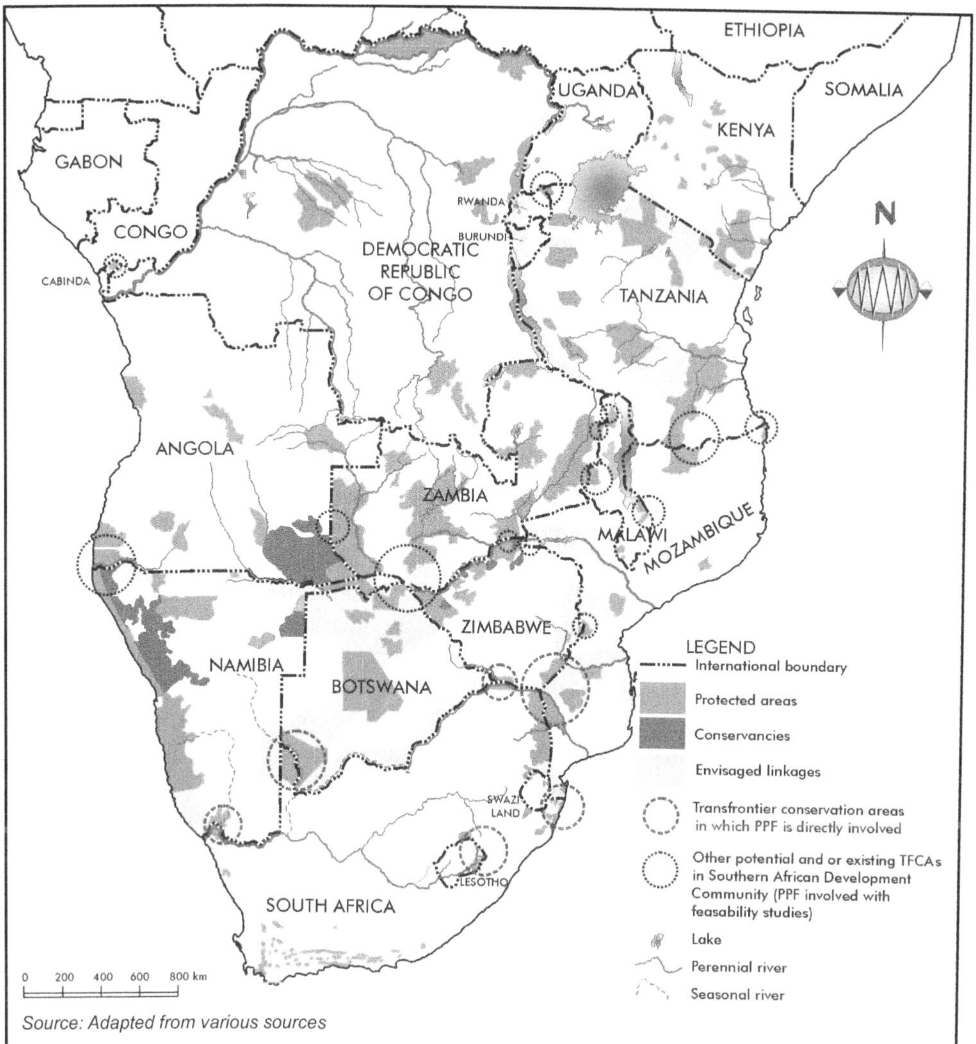

Source: Adapted from various sources

Figure 6.3 TFCAs and the envisaged linkages.

Botswana government has been removing the Basarwa from Xade inside the nature reserve to New Xade, some 70 km outside the western reserve border. This has resulted in reducing a huge settlement of the Basarwa in Xade to a small village, which has been fighting for its survival in the reserve. It is estimated that in 1999 there were approximately 250 people living inside the reserve. Critics have ascribed their removal either to the state's interests in diamond mining in the reserve (Taylor and Mokhawa 2003) or to the oppression of the Basarwa by a dominant Tswana group, particularly the BaNgwato (Gaborone 1998; Mongwe and Tevera 2000). These explanations aside, the position of nature conservationists in the Basarwa saga suggests that there is interest in the preservation of the reserve as part of the Kalahari wilderness. According to Mongwe and Tevera (2000, 80), 'the Kalahari Conservation Society … raised concerns in 1984 that the continued habitation of the [Central Kalahari Game Reserve] by the Basarwa communities was incompatible with the status of the area as a game reserve'. A year later (i.e. 1985) a Fact Finding Mission was convened to investigate land-use options in the reserve and to make recommendations. '[The] government's subsequent decision to relocate residents at serviced settlements outside the [game reserve] went against the mission's recommendation' (Hudson and Murray 1997, 88) that the Basarwa should be allowed to continue to live in the reserve. In terms of the projected TFCA map of the region, the Central Kalahari Nature Reserve forms an important connection between the southern and northern parts of Botswana on the one hand, and between Botswana and its neighbours on the other hand. The development of WMAs in Botswana and their usefulness in the redrawing of the map of the region (i.e. through TFCAs) is evidence of the link between geographical imaginations and existing state policies. As I will show below, communal conservancies in Namibia played a similar role, albeit in a different context.

Realigning land-use options for TFCAs: the case of conservancies in Namibia

Conservancies[24] emerged from the use of private land for nature conservation. According to Langholz and Lassoie (2001), modern private parks gained recognition nearly 40 years ago when it was acknowledged that many nature reserves world-wide were owned by private individuals in diverse forms. Their significance lies in the protection of habitats and species that are under-represented in state protected areas and in that they 'can serve as temporary bulwarks for threatened lands, protecting them until government becomes willing or able to assume responsibility' (Langholz and Lassoie 2001, 1081). In the United States, the Nature Conservancy, which was established in 1951, manages the largest network of private protected areas in that country (Nature Conservancy n.d., 3).

For the colonies, the system of dual land ownership implied that private land was the preserve of the (white) settlers. The establishment of privately-owned nature reserves was therefore racially exclusive. It is in this sense that whites in Namibia had the exclusive privilege of private land ownership. By 1995, approximately 43% of Namibia's land surface was under privately owned commercial farmland (Barnes and De Jager 1995). The granting of rights to use and benefit from wildlife to white landowners in 1967 paved the way for the development of an exclusive white wildlife industry in Namibia. Those rights were consolidated through the passing of the Nature Conservation Ordinance in 1975. Subsequently, the Namibian Ministry of Environment and Tourism registered 148 private nature reserves totaling 0.9% of the total land area of that country in 1995 (Barnard 1998). Moreover, over 10,000 km^2 of farmland had been consolidated into nine commercial conservancies in 1998. These are 'land units managed jointly for resource conservation purposes by multiple landholders, with financial and other benefits shared between them in some way' (Barnard 1998, 45).

It follows that black people in Namibia were alienated from rights to land and other natural resources. With regard to wildlife, the Nature Conservancy Ordinance Amendment Act of 1996 devolved conditional rights over wildlife and tourism to communities. Shumba (1998, 76) summarised the impact of the Act as follows:

> Before the Nature Conservation Amendment Act of 1996, farmers in communal land had no rights over wildlife, in contrast to commercial farmers on freehold land. In communal areas, wildlife remained state property. The Amendment Act of 1996 transferred this right from the state to communities.

The Act set the stage for the evolution of communal conservancies in Namibia. Accordingly, those conservancies were seen as heralding a significant shift in the natural resource policy in that country. In fact, it has been claimed that Namibia's conservancy programme is the most radical of its kind in Africa (Blackie 1999). It should be noted that communal conservancies in Namibia are viewed in official circles as an important step towards local natural resource management. For this reason, they were recorded as some of the landmarks in Namibia's Human Development Report (Namibia 1998), with Turnbull (2002, 73) describing them as 'a leap of faith which has borne fruit in an amazing way'. Furthermore, communal conservancies can be used as areas into which surplus game can be relocated instead of controlling them by means of culling.[25] They are also instrumental in providing additional land to the country's protected areas. In 1998, 41% of Namibia was communal land (Shumba 1998). It is believed that much of the country's wildlife, scenic and cultural heritage is found on communal lands (Jones 1995), and estimations are that communal conservancies will cover approximately 18% of that country in 2030 (Turnbull 2002).

Conceptually, communal conservancies in Namibia manifest the quest for redress, a process which has been captured by the notion of CBNRM. I do not wish to enter into the debate on CBNRM here. Suffice it to say that CBNRM have been interpreted as closest to the radical conceptualisation of a totally community-centred approach to conservation (Hulme and Murphree 1999). As I have intimated above, Blackie (1999) has extended that interpretation to conservancies in Namibia. However, the geography of conservancies in that country is well placed to assist in the realisation of the TFCA map of the region. In the north-western part of the country, communal conservancies form a corridor that usefully connects the Skeleton Coast National Park with Etosha National Park (Figure 6.3). Admittedly, the boundaries of Etosha National Park and environs are ascribed to colonial and apartheid rule in that country.[26] One of the consequences of partitioning the park was that wildlife became highly concentrated, creating serious problems of wildlife/human conflict. Against this background, communal conservancies serve to re-establish the surrounding environment of Etosha to its pre-apartheid state. More importantly, conservancies in Namibia contribute to consolidating areas that fall within the map of TFCAs in southern Africa. 'Many rural communities have been stimulated to form conservancies' (Barnard, 1998), in part because communal land plays the critical role of connecting areas that are within the boundaries of the envisaged TFCA maps. Thus, communal conservancies are sponsored projects, which form an integral feature of the TFCA map of the region. Unsurprisingly, the PPF allocated over R5 million to wildlife and community-based activities in proposed TFCA areas (WildNet Africa 1999).

By using communal conservancies, local populations are easily grafted into the TFCA project – by the lure of ecotourism. In light of this, conservancies of all sorts that have (and some still will be) developed along the boundaries of southern African states should be analysed in terms of their contribution to imagined geography of TFCAs. Put another way, communal conservancies do not exist in isolation, but are increasingly being used as part of the broad strategy for incorporating communal land into TFCAs. This is clear in the thinking of proponents of TFCAs. For example, Braack (2006, 9) described the roles of conservancies in the proposed Kavango–Zambezi TFCA in these words: 'working with our valued partners, Integrated Rural Development and Nature Conservation (IRDNC), we are assisting communities to establish conservancies which *will not only create corridors for the elephants* to access new habitat and other protected areas, but bring very real tourism benefits to these rural people'.[27]

The foregoing discussion highlights the relevance of the debate about the production of scale to the analysis of TFCAs. Admittedly, the discussion presented in this chapter forms part of the growing interest in analysing environmental dimensions and processes in the context of scale. Studies show that there has been a growing attention to environmental NGOs as actors in scalar politics (Dauncey 1988; Brown and Purcell 2005; McCarthy 2005) and that the integration of ecology and human geography has the potential to enhance our understanding of scales and how they work. In addition, attempts are being made to theorise TFCAs as a process of scaling up (Hughes 2005). In this chapter, I have sought to contribute to this growing body of work by arguing for a differentiated process of glocalisation, which recognises the roles played by the state in that process. This is more appropriate for TFCAs, since these cannot be established without the involvement of the state. States are required to sign the necessary legislative documents such as MoUs and treaties. In practice, the need for the joint protection and management of biodiversity (through TFCAs) and fund raising forces the state to scale up its regulatory functions to a regional entity. In the case of southern African states the relevant SADC protocols are instrumental in the process of scaling up the functions of the state. Admittedly, there are differences in how the functions of southern African states are scaled up, because those states have different constitutions from, and processes by, which they develop their legislative instruments. Nonetheless, all states are involved in the scaling-up process of one sort or another in the establishment of TFCAs.

I have also argued that TFCAs manifest particular constructions of scales that disengage local communities, hence those communities are unable to contest the establishment of TFCAs. A dense network of actors operates at the national and regional scales thereby marginalising the local communities from the processes that produce TFCAs. Against this backdrop, TFCAs involve the double rearticulation of political scales that weakens the local communities' demands and struggles for resources. Apart from their marginalisation from the regionalised process that produces TFCAs, the state also weakens the participation of local communities in the TFCA process in three ways. Firstly, governments formulate policies that pre-empt the TFCAs environment. As I have shown in the case of privatisation in South Africa's national parks, the process of privatisation appears to be addressing national issues whereas the outcome of that process feeds directly into TFCA initiatives. Secondly, protagonists of TFCAs appropriate existing state policies to draw their TFCA maps, as in the case of wildlife corridors in Botswana. Thirdly, governments have been encouraged to develop policies that realign land-use options in order to facilitate the drawing of a preconceived map, as attested to by the development of communal conservancies in Namibia.

Against this backdrop, community-based natural resource management projects mediate conflict over natural resources while, at the same time, ensuring that communal land functions as corridors for transfrontier conservation. Conceptually, communal struggles for natural resources are weakened through the reconstitution and restructuring of the local scale as the examples of Zimbabwe have shown. It would be hard, if not impossible, for local communities to oppose TFCAs since they are indirectly participating in the development of those TFCAs. In this case opposition to the TFCA process would imply opposing state policies that feed into that process. Since such policies are couched in the language of redress, they are embraced by local communities, governments, donors and academics as progressive.

Endnotes

[1] Effectively, land was central to liberation struggles in developing countries, and in Africa in particular.

[2] For example, the Landless People's Movement has been formed in post-apartheid South Africa to push for the release of land for the poor.

[3] Community-based natural resources management projects are expression of this belief.

[4] Methanex is a manufacturer of methanol. The company was banned from the State of California in December 2002, because it was said to be polluting the State's groundwater (for details, see McCarthy 2005).

[5] Interview, Werner Myburgh, 5/11/2002, Stellenbosch.

[6] Interview, Deborah Kahatano, 25/8/2003, Gaborone.

[7] There was no way in which Mozambique could have participated in the GLTP initiative without establishing the park. After all, the park is the only area in Mozambique that has been recognised in the Treaty.

[8] The results of the Structural Adjustment Programme in Africa are an example of this.

[9] The former Director of IUCN-ROSA, Katerere, was dismissive of the idea of peace parks (personal communication, 19/10/2000, Bellville).

[10] Personal communication with Carmel Mbizvo, 4/3/2004, Randburg.

[11] Interviews with anonymous high-ranking official, Pretoria.

[12] Personal communication, 5/3/2004, Randburg. Gilberto has a background in wildlife from Wildlife College (Tanzania). He was appointed Park Manager in June 2001, 4 months before the official declaration of the park.

[13] This assumption is based on the centricist tendencies of ruling political parties in the region.

[14] By taking away land held in common.

[15] As in the case of the Kgalagadi and Ais-Ais/Richtersveld transfrontier parks.

[16] Dzingirai (2004) describes the Sengwe as a multi-ethnic group that has long been bypassed by development, and whose economy is organised around border-jumping, poaching, smuggling and so on.

[17] This scenario will require the rezoning of Gonarezhou National park as a wildlife-based multiple use area.

[18] Including the Chitsa community who invaded Gonarezhou National Park to demand the return of their land.

[19] Interview, Johan Verhoef, 2/6/2005, Pretoria.

[20] See Chapter 8 for additional information.

[21] Hughes prefers to call them Euro-Africans.

[22] The airport has the capacity to land Boeing 737, 747 and 767.

[23] I became aware of this description when flying SAA from Cape Town to KMIA in January 2003.

[24] As areas of private property designated for special protection at the initiative of the landowner, in order to conserve the natural resources in the property (Nature Conservancy 2001, 21).

[25] Interview, Dr Fanuel Delmas, 1/7/2003, Windhoek.

[26] For example, the infamous Odendaal Commission led to the removals of the Ovambo from Etosha National Park into 'homelands' in the 1970s. As was the case in South Africa, the Ovambo were removed into their 'homeland'. Consequently, Etosha National Park, which originally covered 93,240 km^2 in 1907, was dramatically reduced by over 70% in 1963 to gain land for the ethnic partitioning under the terms of the Odendaal Commission Report (Lawrie 1963; Barnard 1998).

[27] Other examples of communal conservancies servicing TFCAs include Shewula (Swaziland), which has been used to link up Mozambique, South Africa and Swaziland in the Lubombo TFCA.

Chapter 7

The renaissance of the bush: the reinvention of the Cape to Cairo route in contemporary Africa

Though TFCAs appear to be new ways of reconnecting the African landscape, attempts at connecting different parts of Africa are not new. Politically, the quest for uniting Africa was endorsed by the Organisation of African Unity in 1963. Since then, the ambition for African unity has been central to the Pan-Africanist project and the African renaissance (Nabudere 2001). In contrast to the political goal of African unity that seeks to unite Africa and its people in order to reclaim their respectful position among other nations of the world, colonialism sought to bring different parts of Africa together to create 'regions of control'.[1] The point here is that colonialism aimed at dividing the continent, and at bringing the divided parts of the continent together under the control of a particular imperial power. The same logic of a coherent colonial system was applied at the national level where colonialism sought to bind diverse groups for administrative reasons (Coleman 1994, Mamdani 1996) and to impose the desirability of its own memory above the memories of many African cultures (Mudimbe 1988). The British in particular aimed to promote their influence in, and control over, Africa by linking the territories from the south to the north of the whole continent, in what is commonly known as the Cape to Cairo route.

This chapter examines the Cape to Cairo idea in its colonial and postcolonial contexts. It seeks to excavate the content of that idea in the age of imperialism so as to reassess its re-emergence in 21st century Africa. Initially inspired by commercial interests (Lunn 1992), the Cape to Cairo idea developed into a sharp colonising instrument. Its reappearance as a nature conservation and tourism scheme begs the question of the manner in, and extent to, which colonial ideas are reinvented. This chapter suggests that the construction of the Cape to Cairo idea in different historical moments provides a fecund opening for analysing the colonial presence. It argues that both the British imperialist vision for connecting Africa through the railway and the telegraph, and contemporary trans-Africa eco-based conservation initiatives have common political and economic resonance, and also deploy similar tactics. The fact that southern Africa has been a launching pad for ideas for a continent-wide linkage in different historical moments suggests complex ways in which geography is embroiled in the propagation of visions for Africa and its future.

Trajectories of the Cape to Cairo imaginative geography

Though British imperialism in Africa arose from a constellation of influences, its policy of connecting the south and north of the continent as its terrestrial possession is largely associated with the Cape to Cairo idea. It has in fact been suggested that Britain did not have a clear policy on Africa, and that the evolution of colonies followed a rough path (Coleman 1994). Equally, the Cape to Cairo idea was neither preconceived nor originated as an instrument of imperialism from the onset. Instead, it emerged as a result of the fusion of interests among disparate European agents. The interests of various groups and individuals in Britain[2] coalesced into a common purpose, which found expression in the Cape to Cairo idea. According to Tabor (2003) the first impulse to the idea of an all-British link from the southern to the northern parts of the continent came from the explorer H.M. Stanley's letters to the *Daily Telegraph* in Britain. It is held that the lead writer of the *Daily Telegraph*, Edwin Arnold, expanded the idea and coined the phrase 'Cape to Cairo' in 1876 (Gross 1956; Merrington 2001; Tabor 2003). Subsequently the idea assumed a geopolitical significance.[3] It fascinated British Empire builders and, at the same time, created a climate of rivalry among colonising powers, in particular the French, Germans and Portuguese. As Raphael (1936, 21) explained: 'in the brief caption "from Cape to Cairo," we find an epitome of an important chapter in the history of British expansion in Africa and in the story of international rivalry on the Dark continent'.

Nowhere is Britain's Cape to Cairo expansionist approach more pronounced than in the imperialist Cecil Rhodes's ambition to create and amalgamate contiguous British territories in Africa. In fact, the Cape to Cairo idea is generally considered as Rhodes's vision, and it has been suggested that Rhodes claimed to be its founder (Weinthal 1922). According to Merrington (2001), Rhodes embraced the idea because it expressed all he had ever dreamt of, namely, the establishment of a British World Federation in which a British Pax Africana formed an important part. His universalist vision went beyond narrow British imperialism to encapsulate 'an expression of a totalising concept of history as destiny' (Merrington 2001, 331). The view from Table Mountain in Cape Town (and later from the Matopos in present-day Zimbabwe)[4] was the lens through which he read that destiny. As Rhodes's architect and close associate, Herbert Baker (1934, 164) wrote: 'So Rhodes from Table Mountain dreamed of the extension of the Empire from Cape to Cairo: it was Sea-Quest inspired the one; Land-Quest the other'. Unsurprisingly, mountains and the source of the Nile were important symbols in Rhodes's way from Cape to Cairo (Baker 1934).[5]

It follows that the Cape to Cairo is a geopolitical descriptor of Africa as an important heartland. Indeed, the stalwart of the Cape to Cairo railway, Robert Williams (1922, 149–150), argued that a railway line that connects the whole of Africa from the south to the north, 'portend for Africa, and not for Africa only, but for Asia, Europe and the World', and warned that it would have 'political and economic affects of a very far-reaching character'. Politically, an all-British hinterland from South Africa to Egypt was crucial for Britain's contest for the control of Africa against other colonising powers. It is precisely this political significance that turned the Cape to Cairo idea into a geopolitical strategy for British imperialism in Africa. Before this, the British government was reluctant to endorse the Cape to Cairo idea, with officials variously calling it 'imaginative madness', 'a claptrap', 'sentimental jargon' and a 'curious idea'.[6] However, all these became irrelevant when the British government was made aware of the valuable territory that could fall under British rule were the Cape to Cairo vision to become a reality. Gross (1956, 171) captured how the idea excited British Prime Minister Salisbury in these words:

> Salisbury went to the map on the wall and let his finger slide slowly from the Cape over Bechuanaland, Matabeleland and Mashonaland, crossing over the Zambezi to Nyasaland along Lake Tanganyika, over Uganda and the Sudan, until he stopped at Cairo. In a hushed voice he said: 'Cape to Cairo – Cape to Cairo … and all British.

The Prime Minister's interest in the project was necessary for Rhodes's land quest to the north and his wish to see the hinterland from the Cape to Great Lakes red with British occupation. He was determined that, 'if money is the only hindrance to our striking north from the Zambezi to the headwaters of the Nile, I will find the money![7] ... what was attempted by Alexander, Cambyses and Napoleon, we practical people are going to finish' (Gross 1956, 172). I emphasise that the Cape to Cairo idea was instrumental in the construction of Africa as a heartland. This observation does not invalidate the view that India was the core of Britain's strategy for global power (Hobsbawm 1994). Rather, it is to acknowledge that the Cape to Cairo idea was supportive to that core as both Egypt and the Cape were important short and long sea-routes respectively. Moreover, the promoters of the Cape to Cairo 'saw trading centres along the route developing into colonies, and finally, the colonies merging into a British Empire from Cape to Cairo' (Raphael 1936, 9). The presence of the British in Egypt and the Cape gave Britain an upper hand in the quest for control of areas of strategic importance. For the purpose of our discussion, Britain's presence in those two places, together with the resource endowment in the interior, made the Cape to Cairo idea more appealing to the British public. As Raphael (1936, 21) put it, 'with British interests already strongly entrenched in the south and north: in Cape Colony and Egypt, it was therefore natural that the British public should look with longing eyes at the tempting territory which lay between these two extremes'. The journalist who was formerly in US intelligence, Mark Strage (1973), commented that the Cape to Cairo concept had a geographic logic, which suggested that a master hand was needed to fill the space between the two areas that were already under a firm control of the British, namely, the Cape and Egypt.

The Cape to Cairo idea was cultivated both in Europe and South Africa, the aim being to naturalise the links between the Cape and Egypt on the one hand, and between Europe and the two African countries on the other hand. Merrington (2001) refers to the process of achieving that naturalisation as 'staggered orientalism'. In essence, the process aimed to construct disparate places as parts of the same region in which common identities could be forged. Paasi's (2002, 137) analysis of regions has relevance to staggered orientalism:

> The simplistic assumption of regional identity is that it joins people and regions together, provides people with shared regional values and self-confidence, and ultimately makes the region into a cultural-economic medium in the struggle over resources and power in the broader socio-spatial system.

With regard to the Cape to Cairo idea, different raw materials were mobilised and used in the construction of the Cape as a naturalised ally of Egypt. These included climate, architectural style, population dynamics and travel. Whereas similar climatic conditions in the south and north of the African continent were regarded as the obvious element of the constructed Mediterranean region, architecture had to be mobilised as a force in region formation. Cecil Rhodes's architect, Herbert Baker, was sent to study the masterpieces of architecture and sculpture of Egypt with the purpose of transferring that knowledge to the Cape (*Cape Argus* 1962). The Cape to Cairo connection through architecture was part and parcel of Christopher Wren's doctrine that 'architecture has its political use; it establishes a nation' (cited in Baker 1934, 10). To give the climatologically defined Mediterranean Africa a human face and effect, the Cape Malays were presented as similar to Mediterranean Muslims, the Cape Bushmen to Ur-people in northern Africa or southern Europe such as the Hysos or the Etruscans (Merrington 2001). The lush cultivated lands along the Orange River (now the

Gariep) were likened to the fertile Nile Valley and Afrikaner trekkers from the Cape to the north of the country (i.e. South Africa) thought they had reached the Nile in Egypt when they named the river Nylstroom (Merrington 2001).

Trans-Africa telegraph

Whereas the picturing of the Cape and Egypt helped in the development of the Cape to Cairo consciousness, the continental telegraph, railway line and air routes provided the technical expression of connecting the African continent along the Cape to Cairo axis. They marked areas that were to be subjected to British rule and control. As I will show below, the railway line in particular was useful in the amalgamation of British territories, and together with the telegraph and air route (along the Cape to Cairo axis) defined Britain's geopolitical position on the African continent. The frontrunner of the railway system was the trans-Africa telegraph,[8] which was meant to facilitate the administration of colonies and to facilitate commercial transactions. The telegraph project became public in 1878, when the Cape Government offered a subsidy of £15,000 per annum for 15 years to any person or company who would construct and work a telegraph between England and the Cape (Weinthal 1922). Cecil Rhodes looked upon it as, 'a kind of advance guard of civilisation,[9] which would, in due course, be followed by a railway system' (Weinthal 1922, 217). The suggestion for an 'all-British telegraphic cable' had a peculiar fascination among Empire builders, who looked upon it as 'an all-powerful weapon for developing trade between the Mother Country and her Colonies, and for bringing about something in the nature of Imperial Federation or a Pan-Britannic Customs Union' (Bright 1911, 35). It was proposed that the telegraph line should be as important as the British navy and army, and that it should be a matter of the state rather than the private sector to enable the British to talk freely 'yet secretly with the rest of the Empire' (Bright 1911, xvi). Bright (1911) was worried that the inter-imperial news to and from the United Kingdom was being subjected to the control of American interests and party policies in the absence of a British-controlled cable. He viewed the all-British telegraph between the United Kingdom and Canada as an example to be followed by the British government and suggested that the telegraph line in Africa should follow a projected all-British steamship. In the context of the Cape to Cairo, the all-British Atlantic cable was to serve as a second and superior strategic line to the Cape. In this sense, the telegraph set the scene for the Cape to Cairo connections. However, it was the railway that played a crucial part in the realisation of those connections. Strage (1973) has suggested that Rhodes's Oxford classmate, Charles Metcalfe, was probably the first person to put the idea of the railway running the length of Africa in print, and who also had a proposal for the whole plan.

The railway: servicing the Cape to Cairo

The development of the railway in South Africa was spurred by the inauguration of the Stockton to Darlington Railway, which was opened in England in 1825 (Hoare 1975). Subsequently, the Cape Colony Government proposed building a railway in 1828 (Strage 1973). The Provincial Committee of the Cape of Good Hope Western Railway considered building a railway in Cape Town, and the first Railway Act of the Cape was passed in 1857. This opened the way for the Edward Pickering (company) to contract to build the railway line from Cape Town, tapping the corn and wine districts of Stellenbosch and Paarl, to Wellington in 1858 (Raphael 1936; Hoare 1975). In the same period, Britain was pursuing the development of the railway between Alexandra and Cairo, and Cairo and the Suez in Egypt. These developments laid the foundation for both the southern and northern sections of the transcontinental railway. The sections of the railway at both ends of the continent clearly marked the path along which the continent of Africa could be connected, and stimulated the imperial imaginations of the natural resources that could be brought under British control. In Rhodes's imperial vision, the Cape to Cairo railway became 'the sturdy armature on which an empire could be assembled, shaped and cast' (Strage 1973, 16).

Africa's minerals played a critical role in the development of the Cape to Cairo imperial mission. The discovery of gold on the Witwatersrand led to a belief that similar goldfields lay beyond the Limpopo (Malcolm 1939), hence Matabeleland became a target. For the purpose of our discussion, mineral prospecting accentuated the desire to link the south and north of the African continent. This became clear when Rhodes consolidated mining companies under the aegis of the British South Africa Company (hereafter the Chartered Company). In its application for the Royal Charter, the Company presented its main feature as 'to extend the Cape railway and telegraph systems to the Zambezi River, to encourage British colonisation, and promote trade and commerce in the districts thus opened up, and to develop the mineral and other concessions which they already possessed' (Weinthal 1922, 412). The Colonial Office wanted to limit the operation of the Chartered Company between the Cape and Zambezi, but Rhodes used the Cape to Cairo *leitmotif* in his argument for the permission to operate beyond the Zambezi. Consequently when the company was incorporated by Royal Charter on 29 October 1889 (Malcolm 1939), 'the Charter contained no mention of the northern boundaries in regard to the extension of the Company's territory', and Rhodes regarded the concession given to his Chartered Company 'so gigantic, [it was] like giving a man the whole of Australia' (Gross 1956, 181).

More applicable to the Cape to Cairo vision is that the Charter was useful in two main ways. Firstly, by opening up vast lands to the north of Limpopo, the Charter embodied the territorial indefiniteness and legalised Rhodes's northwards expansion. The De Beers Trust Deed, as drafted by Rhodes, empowered the company to 'acquire any asset it pleased, including tracts of country in Africa and elsewhere ... to build and operate tramways, railways, roads, ...' (Strage 1973, 46). Secondly, it encouraged expeditions along the Cape to Cairo axis. For instance, the expeditions of 1899 and 1901 were sent to Africa for the purpose of 'discovering a mineral area of sufficient importance to draw the line northwards' (Williams 1922, 148). In this sense, mineral discoveries were used as justification for the extension of railways, while areas rich in minerals fell into the hands of imperial entrepreneurs.[10] Rhodesia was the first main target, because it provided the strongest, yet highly contested, links in connecting the Cape with the northern territories. In his foreword to Weinthal's (1922, xxvii) four volumes, Jan Smuts described the role of Rhodesia in the Cape to Cairo vision in these words: 'South Africa has been most intimately connected ... with the conception of the Cape to Cairo route ... [It] was [Rhodes's] great vision, and the land called after his name [i.e. Rhodesia] was his creation, and incidentally the greatest step forward towards the realisation of the vision'.

The attempt to incorporate Rhodesia into the Union of South Africa failed following the Referendum of 27 October 1922, in which Rhodesians decided in favour of a Responsible Government by 8774 votes to 5989 (Gale 1935). Moreover, the Afrikaner-dominated South African Republic made that incorporation geographically impossible. Under these political conditions, Rhodes regarded Bechuanaland (present-day Botswana) as the gateway to the north, hence he persuaded the reluctant British government[11] to colonise Botswana in order to, among other things, keep his northward expansionist plan intact. Bechuanaland performed the same strategic role that the Suez Canal played in the geopolitics of Europe, hence that country has been dubbed the Suez in the description of Britain's northwards expansionism in Africa. The relationship between the British Government and the Chartered Company over Bechuanaland was sealed by the tripartite agreement[12] between the Government of British Bechuanaland, the Chartered Company and the Bechuanaland Railway Company, which was signed by the three parties on 9 August 1894 (House of Commons 1894). It should be emphasised that the colonisation of Bechuanaland by Britain was an important step in the development of the road to the North. As I will be showing below, both Botswana and Zimbabwe still remain strategic territories in the newly relaunched Cape to Cairo rhetoric. For the moment I highlight the practices that bolstered the Cape to Cairo imperial vision. Such practices are important for understanding how geopolitical ideas are translated into everyday life. As Dodds (2001, 474) noted, 'critical geopolitics needs to engage with … concerns over the role of practices in everyday life'.

The colonial romance of the Cape to Cairo route

> 'Speak of Africa and golden joys'; the joy of wandering through the lonely lands; the joy of hunting the mighty and terrible lords of the wilderness, the cunning, the wary, and the grain. In these greatest of the world's great hunting-grounds there are mountain-peaks whose snows are dazzling under the equatorial sun; swamps where the slime oozes and bubbles and festers in the steaming heat; lakes like seas; skies that burn above deserts where the iron desolation is shrouded from view by the wavering mockery of mirage; vast grassy palms and thorn-trees fringe the dwindling streams; mighty rivers rushing out of the heart of the continent through the sadness of endless marshes; forests of gorgeous beauty, where death broods in the dark and silent depths (Theodore Roosevelt, cited in Weinthal 1922).

Apart from mineral prospecting, the propagation of the Cape to Cairo idea was bolstered by a number of activities, the most important one being travel. It should be acknowledged that travels to Africa predate the Cape to Cairo invention. Nonetheless, the advent of the Cape to Cairo railway promoted a particular pattern of travel that naturalised the route. Arguably, those who supported the construction of the railway line had interest in trans-border tourism. For example, Jan Smuts commented that:

> when the engineer has completed his task something will still remain beyond the range of mechanics. Africa will remain the unique, the indefinable, the most baffling of the continents. But its fascination will no longer be confined to the select circle of great travellers and discoverers and mighty hunters. The route will make the great secret of Africa the possession of all who care to travel (cited in Weinthal 1922, 28–29).

Of relevance to the discussion of this chapter and the book is that the Cape to Cairo, as an idea as well as the route, realigned continental travels in and over Africa (see Grogan and Sharp 1900; Letcher 1905). Such travels were considered of imperial value. For example, Jan Smuts encouraged Stella Treatt and her husband to embark on a Cape to Cairo trip and to write about that trip for people in England so as to create interest in the Cape to Cairo expedition (Treatt 1927).[13] Their route stuck to the all-British territory and they received assistance from the colonial office (Treatt 1927). Belcher (1932) reported about similar experiences of two women who organised a trip from Cape Town to Cowley via Cairo, in which a broad red line was clearly marked from Cape Town to Nairobi, and from there continued almost without a break to Cairo. In the case where the journey did not begin from Cape Town, there was an expressed interest among travellers to join the Cape to Cairo route, as it was the case with Captains Speke and Grant. Their route across the continent commenced in Zanzibar, but it joined the Cape to Cairo route northward from Lake Victoria (Grant 1864).

Leo Weinthal founded the Cape-to-Cairo Express and promoted tourism in Egypt. Robert Murray, in conjunction with Thomas Cook's travel business explored tourism between Cape Town and East Africa, the aim being to show tourists 'the glories of unexplored Africa' (Murray 1996, 29). Subsequently, Murray formed the Trans-Africa Safari Company[14] 'to operate fully inclusive tours everywhere in Africa between Cape Town and Juba[15] ... and his drivers wore dark blue uniforms and peaked caps with a silver badge in the shape of a map of Africa' (Murray 1996, 36).

Air transport was also used to support tourism along the Cape to Cairo axis. For example, Air Vice-Marshal Salmond hoped to establish Cairo as a centre of three great empire air routes to India, to the Cape and to Australia (Weinthal 1922). The Cape to Cairo air route opened on 27 December 1919 (Weinthal 1922). According to the Britannia website,[16] nowhere was the romance of commercial flying more compounded with the perceived conditions of the exotic than on the Imperial Airways[17] route between Cairo and the Cape. The African route was considered the world's most adventurous sky journey.

Travels on the Cape to Cairo route also conferred a sense of achievement. Mary Hall's (1907, v) preface to her book makes this point clear: 'as I am the first woman of any nationality to have accomplished the entire journey from the Cape to Cairo, I think perhaps a simple account of how I managed to do it quite alone may be of some interest to many who, for various reasons, real or imaginary, are unable to go so far afield'. Belcher (1932) suggested that using a light car, the Oxford Morris, for travelling from Cape to Cairo demonstrated what British goods could do. The same sense of achievement and the spatiality on which it is embedded is manifested in 21st century Africa, as I will show in the section that follows.

Transfrontier conservation: the Cape to Cairo by another name

Despite its imperialist impulses and connotations, the phrase 'Cape to Cairo' has become a descriptor of projects that are supposedly meant to benefit post-independence Africa. It is being used in contemporary Africa on the grounds that its purpose and meanings could be rehabilitated to fit into current social, political and economic realities. Put it differently, it is believed that the Cape to Cairo axis could serve a purpose different from its imperialist one and that there is value in recycling that phrase under a different political environment. To this end, Naidoo[18] (cited in Banfield 1999) suggested that the information super-highway should be built from Cape Town to Cairo to liberate the continent, arguing that the same Cape to Cairo axis that had been used to subjugate the continent could also be an instrument for its economic liberation. In the same vein Robertson (1999)[19] described the signing of an agreement for the rail link from Johannesburg to Kampala in 1999 as the rail deal that links Cape to Cairo.[20] The message here is that the Cape to Cairo axis is as relevant for present-day Africa's development as it was claimed to be in the past. The imperial motives behind the expansion and connection of British territorial possessions in Africa are being reinterpreted in 21st century Africa in order to promote tourism. In the context of tourism, the Cape to

Cairo is viewed as presenting an opportunity for the tourism industry to harness Africa's geographical heritage (Nelson 1996). Accordingly, Frost and Shanka (2001) have argued for the revival of the Cape to Cairo dream for tourism purposes.[21] They consider the Cape to Cairo axis as a corridor that is viable for tourism from both the political and economic standpoints. Politically, the corridor is viable for tourism because 'African governments no longer view international tourism investors as agents of imperialism or neo-colonialism', the tourism route along the corridor 'has strong social and economic consequences for the participating nations' and Africa has entered a new political climate of co-operation and regionalism (Frost and Shanka 2001, 236). Economically, the corridor has a well-developed infrastructure with countries in the corridor offering viable tourist products to the existing international market. Using the tourism data from 1989 to 1998, the authors argued that the corridor accounts for an average 52% of all arrivals in Africa.

It follows that tourism is been promoted on the historical idea of the Cape to Cairo, the aim being to relive the colonial Cape to Cairo journeys and experiences that have, as the *National Geographic* explorer, Holgate (2003), laments, sadly been lost. That is to say that the Cape to Cairo tourism routes capitalise (Figure 7.1) on historical amnesia, promoting the quest to follow on the footsteps of European explorers. For example, Figenschou wrote on the VW website that 'ever since I learned as a schoolboy about the British imperialist Cecil John Rhodes and his vision of the Great North Road from Cape to Cairo, I have dreamed of driving the length of Africa, earth's second largest continent'.[22] Like in the days of Rhodes, most journeys across Africa start from Cape Town, with South Africa acting as a transition zone between the civilised north and the uncivilised Africa. As the EcoAfrica website put it, 'South Africa is the first country you will encounter on this journey [i.e. Cape to Cairo]. It's a civilised introduction to the vast and untamed continent that is Africa'.[23] The spirit of conquering the historically perceived 'dark' and 'hazardous' continent is also being nurtured. In her account of Brian Routledge's retirement journey from Cape Town to Cairo, Redfern (1999) alluded to the view that one can travel across Africa by car if one has 'the right equipment, time and nerve enough to tackle the hazards of a continent struggling with third-world social and political problems …'.

Different modes of travel are being used to promote travelling along that route, the main ones being air and road travel.[24] For example, a Washington-based travel company, Cape to Cairo LLC, described itself as a tour operator, airline consolidation, and safari specialist for Southern and East Africa that provides complete individual and group travel planning for African safari, vacation and business travel.[25] Tourism Africa also specialises in Cape to Cairo air safaris. It organised such safaris on 3–17 November 2004 and on 2–16 March 2005 at a cost of US$12,000 per person sharing. Similarly, the Trans-Africa Expedition by a Malaysian expedition team travelled from Cape Town through Namibia, Botswana, Zimbabwe, Zambia, Malawi, Tanzania, Uganda, Kenya, Ethiopia and Sudan to Cairo, Egypt, in 2000 (Frost and Shanka 2001).

In southern Africa, overland travel on the Cape to Cairo route, especially by four-wheel drive (4x4s), has gained momentum. For example, the 4x4 Eco-Challenge, which was first organised in 2001, was an overland adventure 'geared to the concept of enjoying the untamed beauty nature has to offer' (*Drive Out* 2005, 8), and was considered an overland tour for testing co-operation, knowledge and skills. Admittedly, not all 4x4 overland tours follow the Cape to Cairo route. Our main concern here is with tours on the Cape to Cairo route and with those that are supportive to, or emerge as a result of, the peace parks initiative. The 4x4 makes crossing borders easy, as the *Drive Out* magazine suggests. In 2002 *Getaway* magazine organised an overland Cape to Cairo trip, calling it 'the quintessential African

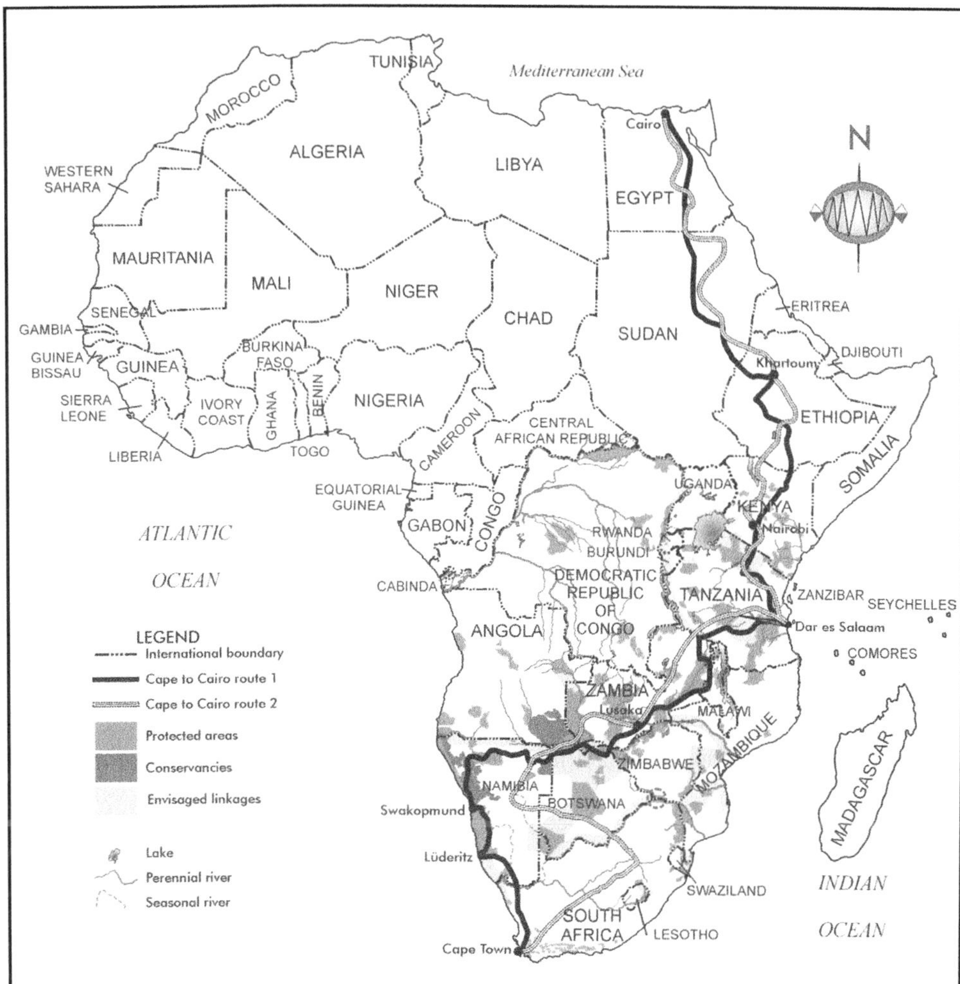

Source: Compiled from various sources

Figure 7.1 Cape to Cairo routes and TFCAs.

Journey' (*Getaway* 2002, 126) under the guidance of an 'experienced African nomad', Mike Copeland.[26] The cost of the trip was later calculated at R43,653. The outcome of the journey and the involvement of *Getaway* as the organiser of that journey have relevance to the theme of this chapter and, indeed, the volume as a whole. The ultimate goal of the journey was to research material for *Getaway*'s Cape-to-Cairo guidebook that was published in 2003. The first two paragraphs of the introduction to the guidebook have these words: 'we are all nomads at heart and Africa is the ideal place to follow your instincts … If it is adventure and danger you crave, then it's here in truckloads. The mountains, jungles and deserts will fill you with awe … wild animals roam the open spaces like bit players in a timeless saga …' (Copeland 2003). The guidebook does not only emphasise attributes of the African landscape that appeal to tourists, but also provides practical steps towards travelling across the continent.

TFCAs and tourism

As we have seen in the previous chapters, TFCAs are promoted primarily on ecological grounds, and are also couched in the language of sustainable development. The tourist industry has taken the ecological and sustainable development agenda further by promoting tourism as an appropriate economic activity that would benefit nature conservation and local communities. Notably, the expansion of national parks in the form of transfrontier parks is helpful in recreating the wilderness, which the tourist industry could sell as a (tourism) product. Unsurprisingly, TFCAs have the full support of the tourism industry and 4x4 trails are marketed as one of the attractions of peace parks. Both the Great Limpopo, Kgalagadi and Ais-Ais/Richtersveld peace parks have self-drive trails (*Drive Out* 2005).

It could be argued that TFCAs do not only create the wilderness for tourists, but also contribute to the revival and popularisation of the imperial-inscribed Cape to Cairo route. Evidence of this is found in comments about TFCAs and the Cape to Cairo idea and, more importantly, from the alignment of contemporary tourism routes, in which emphasis is placed on the routes from Cape Town to the interior of the African continent. For example, Addison (1997, 66) drew the parallel between Cecil Rhodes and peace parks when he commented that:

> a hundred years ago Cecil Rhodes impudently … announced his vision of an Africa painted red under the British flag. Now under the rubric of peace parks, the map is to be coloured green from Cape to Cairo with the support of African presidents and major funding from private sector investors and donors in the wealthier countries.

Getaway supports the creation of peace parks, which it sees as a grand ecotourism plan. As the *Getaway* writer Pinnock (1996, 89) put it, 'the relinking of these ancient corridors [through peace parks] could re-establish the great herds … and provide the basis for a huge ecotourism industry'. In fact one of the ways the maps of the plan for connecting TFCAs in southern Africa was first brought to public attention was their publication by *Getaway* magazine in November 1996. The 'superparks', as *Getaway* calls them, were to begin from existing protected areas.

It should be noted that *Getaway* is not alone in promoting the idea of ecotourism routes from Cape Town to the north of the continent. The Cape Town-based Open Africa Initiative, which was established in 1995 by a group of scientists, business people and tourism experts, embarked on the 'African Dream Project' that aimed at linking the splendours of Africa through a network of Afrikatourism routes from Cape to Cairo. The idea is credited to the founder of the African Dream Project, Noel de Villiers. It should be noted that the African Dream Initiative developed alongside that of cross-border parks – a process that has led to the interpretations of the African Dream as both a nature conservation and tourism project. It has been speculated that the African Dream Initiative was overtaken by the development of peace parks, which gave the Initiative a geographical expression. Addison (1997) is of the view that it was Anton Rupert who added the final mix to the African Dream through the promotion of peace parks. The manner in which the idea of connecting Africa by means of green belts of protected areas was created and carried out could be debatable, but the link between peace parks and the African Dream Project is identifiable. The Project painted the image of Africa in these words:

> Indulge in some imaginary scenario-building. Redefine 'game parks' to include areas reflecting the traditions, culture, arts, literature, archaeological sites, and folklore of Africa. Imagine these linked from Cape Town to Cairo, such parks to serve as the cornerstones of an African paradigm and modality of living that harmonises with nature. Pencil in accommodation and new forms of craft and other industries along these links and on the borders of these parks, with informal markets employing hundreds of thousands of people; agriculture based on farming with wildlife; music, clothing, art, and theatrical exports with nature as the theme, education with an emphasis on environmental management; new forms of architecture based on old. See Africa vividly etched against the background of a worsening environmental crisis almost everywhere else on the globe. Anticipate the human interest in nature 30 years from now to be far greater than can be imagined today, and the yearning for wilderness areas to have intensified by then. Regard Africa as the world's garden, the globe's lung, as the place from which humanity derives a new ethos and ethic environmental inspiration (De Villiers 1995, 9–10).

De Villiers thought that the image of Africa that he painted above would galvanise moral and financial support for the project, more especially in South Africa where it was launched.[27] Consequently, he underestimated the need for fund-raising, which has been the main drawback for the project. Ten years later (i.e. 2005) the Nelson Mandela-patroned African Dream Project was still operating mainly inside South Africa. It launched its first route in 1999 and had 56 routes in seven countries by 2005. The project's first route outside South Africa was opened in 2002 in Swaziland in the Lubombo region (Rodrigues 2002).[28] It was envisaged that the routes will ultimately link up, acting as forerunners to peace parks. Clearly, the way in which the African Dream Project establishes tourism routes accounts for the localisation of those routes. The routes are developed by a group of people who come together to use features such as culture, historical sites and so on, to promote tourism. Following the identification of the route, Open Africa assists with the planning of the route.[29] According to De Villiers, that approach is modelled on the famous wine routes of the Cape, which were established by wine farmers who were keen to attract tourists by developing activities that contribute to economies of scale.[30] In essence, the voluntary approach by which Afrikatourism routes are developed means that the pattern of the routes is neither predetermined nor enforced. Consequently, the routes are not necessarily aligned to the Cape to Cairo axis. Nevertheless, the Cape to Cairo imagery has been used to encourage the development of tourism routes with a long term view of creating a network of routes across the African continent.

I emphasise that there is a concerted effort towards developing a network of tourism routes that connects with TFCAs in order to promote tourism at the regional level. This is made possible by realigning road infrastructure with existing and potential TFCAs sites, as is the case with the Lubombo route[31] that was launched in May 2006. That route, which forms part of the Maputo Development Corridor, is aimed at the development of tourism in the area. It is anticipated that the route would offer a diversified tourism product that would simultaneously benefit Mozambique, South Africa and Swaziland. Tourists will be able to explore the diverse cultures, landscape, wildlife and the coast. Accordingly, the joint statement of the ministers of the three countries claimed that the route will turn the Lubombo region into 'one of the few places in the world where tourists will be able to dive on coral reefs and observe whales and dolphins in the sea within just a few hours of experiencing a big five game safari'.[32] It should be noted that the Lubombo route is a network of tourism

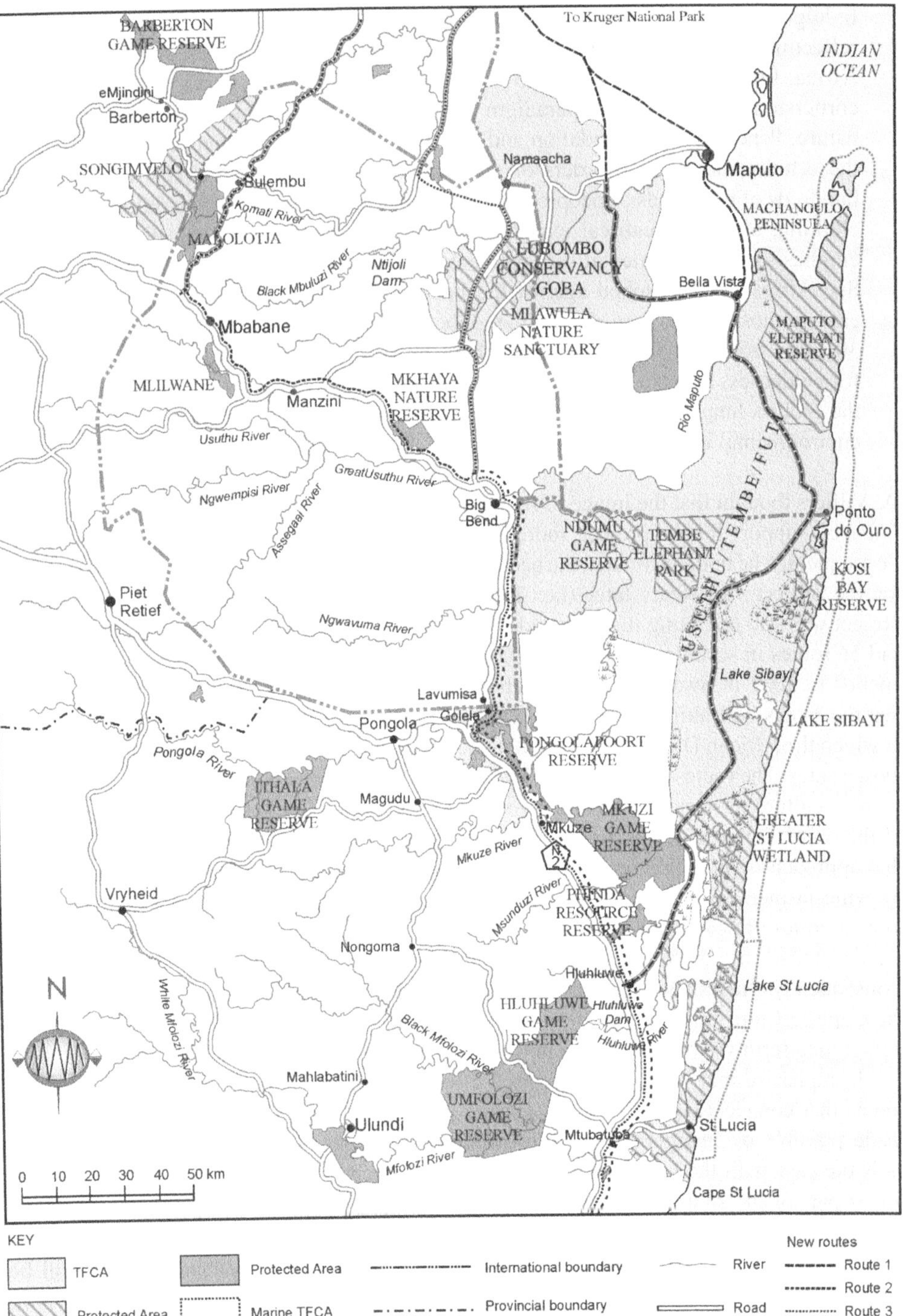

Source: Compiled from various sources

Figure 7.2 Lubombo TFCAs and tourism routes.

routes, namely, the Coastal Adventure Route (Route 1), the Mountain Route (Route 2) and the Ezulwini Valley Route (Route 3) (Figure 7.2). The three routes are all connected to the TFCAs in Lubombo, with Route 1 linking the Lubombo and the GLTP (see Figure 7.2). According to the *Swazi Observer Online*, the Ezulwini Valley Route has been developed to allow visitors to experience the Songimvelo–Malolotja TFCA.

The links between tourism and TFCAs in the region have institutional support in the form of SADC and the Regional Tourism Organisation of Southern Africa (RETOSA). As we have seen in Chapter 5, SADC supports the development of TFCAs in southern Africa. The member states[33] of SADC signed the SADC Charter of RETOSA[34] in 1997, which means that SADC has entrusted RETOSA with regional tourism matters. According to the Charter, the objectives of RETOSA are to encourage and facilitate the movement and flow of tourists in the region, applying the necessary regional or national policies and mechanisms, which facilitate the liberalisation of exchange control and regulations; facilitate a community and rural-based tourism industry and culture throughout the region; and to develop, co-ordinate and facilitate tourism marketing and related promotional opportunities into the region (RETOSA 1998). Of relevance to the discussion of this chapter is that RETOSA's marketing activities are primarily focused on scenic beauty, wildlife, culture and seaside. The organisation's image of the southern African region is that of 'a scenic tropical paradise with beautiful wild open spaces and African animals, backed up by its people's friendliness, good food and service' (RETOSA 1998, 15). Accordingly, RETOSA has branded southern Africa as 'the essence of Africa – beyond your wildest dream', and its logo has four colours[35] that reflect the region's 'African-ness'. It should be noted that the SADC Council approved the recommendation that RETOSA should remain a subsidiary of the organisation of SADC in Mauritius in August 2004 (SADC 2005). The view that TFCAs are harbingers of peace has elevated the link between TFCAs and tourism beyond the southern African region to make them relevant to the entire African continent.

Touring Africa for peace or peaceful tourism in Africa?

Undoubtedly, peace parks play a critical role in the promotion of tourism in Africa, mainly because they support and entrench the romantic view of African landscapes while, at the same time, being aligned with ideas about peace on the continent. There is a strongly held view that Africa has a rich natural and cultural heritage that could be harnessed for the promotion of tourism, which acts as a catalyst for peace in the continent (Sumaye 2003). As on other continents, the IIPT is at the forefront of promoting the link between peace parks and tourism in Africa. Accordingly, it held its first African Conference on Peace through Tourism in Nelspruit, South Africa, in 2002 under the theme 'Building Bridges of Peace through Sustainable Community Tourism Development'. The aim of the conference was to foster African awareness and commitment to the social, cultural, environmental and economic benefits of tourism and the role of tourism as a catalyst for reconciliation and peace on the continent. Subsequently, the Second IIPT African Conference was held in Dar es Salaam, Tanzania, on 7–12 December 2003 (IIPT 2003a), the aim being to develop partnerships, implementation strategies and action initiatives that contribute to sustainable community development, peace and poverty reduction.[36] Significantly, the conference led to the development of the Tanzania Action Plan with a clear agenda that included developing a global awareness of the activities of the IIPT.

To that end, it endorsed the United Nations declaration of the first decade of the 21st century (i.e. 2001–2010) as the International Decade for a Culture of Peace and Non-Violence for the Children of the World. It is claimed that the Conference was actually organised to support that UN declaration (IIPT 2003b). One of the key Action Agendas adopted at the Tanzania conference was sustainable tourism development, which aimed to encourage all nations of Africa to develop sustainable tourism strategies, policies and master plans. More importantly, the Action Agenda encouraged regional collaboration in natural resource and environmental planning consistent with regional ecological systems, and the development of transnational peace parks (IIPT 2003b). In emphasising the need for developing peace parks in Africa, the Deputy Secretary-General of the World Tourism Organisation, Dawid de Villiers (2003, 3), told delegates[37] that 'one of the most successful and powerful initiatives to utilise the immense natural potential of Africa – its open spaces, rivers, mountains and wildlife – comes from the Peace Parks Foundation and the concept of Trans-frontier Conservation Areas'. This is evidence that tourism organisations support the development of peace parks in Africa for tourism purposes. Investors have also come on board as they see travel and tourism as a key factor of the African economy. According to the founder and President of the IIPT, Louis d'Amore (2003), members of the US-based Corporate Council on Africa, which gives high regard to travel and tourism, account for 85% of all US investment in Africa.

The Third IIPT African Peace Through Tourism Conference,[38] held in Zambia on 6–11 February 2005, also has direct relevance to the peace park concept and to the legacy of British occupation in Africa. The goals of the Conference were to: (i) provide a forum for leading government, industry, donor agency, NGO, educators and practitioners to debate and reach consensus on a 21st Century Vision for African Tourism; (ii) facilitate the development of strategic alliances to achieve that vision; (iii) demonstrate elements of that vision with 'success stories' and models of 'best practice'; and (iv) continue building a global awareness of Africa as a 21st century destination (IIPT 2005a). The conference coincided with Zambia's 'National Peace Through Tourism Week',[39] during which Zambia celebrated the 150th anniversary of Livingstone's sighting of Victoria Falls, the Centenary Year of the Town of Livingstone and the 100th birthday of the former UN Secretary General Dag Hammarskjöld. Significantly, Zambia dedicated Victoria Falls as an IIPT International Peace Park. An important outcome of the Zambia Conference was the adoption of the Lusaka Declaration, which called on governments, the private sector and civil society to support tourism based on a unique African experience – the branding of Africa. Moreover, the Lusaka Declaration emphasised the need for regional and sub-regional co-operation and practically supported the concept of peace parks by establishing the Great Falls Peace Parks Initiative, which dedicated IIPT International Peace Parks in each country adjacent to the world's three great falls, namely, Iguaçu (between Argentina and Brazil), Niagara (Canada and the USA) and Victoria (Zambia and Zimbabwe) (IIPT 2005b).

In summary, the imperialist vision for a continent-wide connectivity is being rehabilitated for tourism purposes, with TFCAs playing a critical role in that process. The Cape to Cairo concept and TFCAs have some similarities, which range from their geographical reach through to their impact on Africa and its people. Geographically, both the Cape to Cairo concept and TFCAs emphasise South Africa as a starting point for continent-wide projects of different kinds. They both emphasise the connectivity of countries for the development of the continent. To be sure, the development of the railway across Africa was viewed by imperialists as an instrument for civilising and developing the supposedly dark and barbaric Africa. Tabor (2003) has actually suggested that parts of the continent, such as the Democratic Republic of the Congo, are troubled hotspots, because they were not touched by the railway. He maintains that 'it is ironic that the gaps[40] of less than 500 miles lie in two of the most troubled territories on the continent' (Tabor 2003, v). A similar message is being spread that African countries that do not participate in TFCAs will not benefit from the economic spin-offs of ecotourism and from the 'peace' that TFCAs would foster. As in the case of the imperial Cape to Cairo project, non-state actors, particularly the private sector, are the main drivers of TFCAs in Africa. This begs the question of the extent to which Africa's fortunes and misfortunes are continuously shaped by non-state actors, of course, with the complicity of the state. It follows that touring from the Cape to Cairo is becoming infectious in contemporary Africa as Fox (2004) has suggested, and new titles on the Cape to Cairo route have become popular among book publishers (Hugon 1993; Oberholzer 1996; Holgate 2003; Moore 2003; Theroux 2003). Rhodes's Cape to Cairo ambition, and his ecological imperialism that was manifested by birds and imported exotic fauna (Maylam 2005), is being replayed in the form of TFCAs and their associated tourism.

Endnotes

1 Countries that an imperial power had assembled under its control. Prior to, and immediately following the Second World War, African countries were classified according to the imperial power exercising control.

2 The Cape to Cairo imperialists included Sir John Kirk, Sir Harry Johnston, Captains Lugard and Kitchener and Cecil Rhodes, and were sponsored by influential business groups in England (Raphael 1936).

3 The idea was conveyed to Lord Salisbury by Harry Johnston, who wrote about it from Central Africa (Tabor 2003). Merrington (2001) has suggested that the general policy of extending British control across Africa from south to north was delineated in 1881.

4 See Ranger (1999).

5 Baker, who has been described by Jan Smuts as an artist with deep sympathies and imaginative insight, claimed that Rhodes's imperial vision was a manifestation of Ruskin's eloquent appeal to the Youth of England to go forth and 'found colonies as fast as she is able … seizing every piece of fruitful waste land she can set her foot on and teaching her colonists that their aim is to advance the power of England by land and sea [and] to make England a centre of peace, mistress of learning and of the arts' (Baker 1934, 11).

6 Wills (n.d) expressed the objection to the Cape to Cairo railway as follows: 'a direct Bulawayo–Tanganyika railway would be so utterly useless, and such a hopeless loss to anyone who had to keep it open for traffic … It is more likely that [those who propose the idea] are insincere than insane'.

7 Rhodes believed that one needs money to carry out big dreams (Green 1936).

8 According to Tabor (2003), the African transcontinental telegraph was a brainchild of the Aberdeen-born engineer, Sivewright. Individuals such as Samuel Baker were opposed to the idea of the telegraph, with Baker arguing that the Sudanese will use the wire to make bracelets and arms in these words: 'I do not think any police supervision would protect a wire of gold from London to Inverness and I think it would be equally impossible to protect a wire of iron through the tribes' (cited in Weithal 1922, 214–215).

9 On receiving an unsolicited cheque of £10,000 from Abe Bailey to be used for the telegraph line, Rhodes suggested that a memorial, on which the first subscriber to the Trans-continental Telegraph Company should be inscribed, should be built in South Africa.

[10] According to Malcolm (1939), the traditional agent of imperial expansion was the gold-seeker and merchant, and adventurer.

[11] In 1882, the Colonial Secretary, Lord Derby, told the British House of Commons that Bechuanaland had no value to Britain and that it was of no consequence who owned it (Strage 1973).

[12] The agreement contained the following three main points: (i) for the railway line from Vryburg to Mafeking, the colonial government will make certain land concessions; (ii) for the line from Mafeking to Gaberone, the colonial government will pay a subsidy of £10,000 per annum for 10 years while Chartered Company will pay a subsidy of £5000 per annum for 10 years; and (iii) from Gaberone to Palapye, the same funding arrangements will apply (House of Commons 1894).

[13] Treatt dedicated her book to 'comrades of the long trek'. She suggested that travelling on the Cape to Cairo route was one of the best promises a young man could make to his newly wed wife.

[14] The Company was formed after Murray's first visit to North America in 1948 (Murray 1996).

[15] The southern limit of navigation of the Nile.

[16] www.britannia.com

[17] Imperial Airways was later renamed British Airways.

[18] The former South African Minister for Posts, Telecommunications and Broadcasting. Naidoo participated in the African Rally, which aimed at mobilising funds for building the information network that would help the continent to participate effectively in, and benefit from, the global information technology.

[19] www.btimes.co.za

[20] In practice, the railway will be run by Tanzanian-registered Trans-Africa Railway Corporation.

[21] They maintain that 'the time is opportune to revisit that dream of the 18th century [i.e. Cecil John Rhodes' imperial ambition] – not from a political standpoint, but from a country commercial tourism approach' (Frost and Shanka 2001, 237).

[22] www.vwtrendsweb.com/features/0308vwt_kombi/index.html

[23] www.ecoafrica.com

[24] These are by no means the only modes of travel along the route. For instance, David Duncan (1989) used his mountain bike to travel the same route.

[25] www.capecairo.com (accessed 14/3/2005).

[26] *Getaway* readers were promised a Land Rover Defender Td5 (valued at R291,000) as a top prize at the end of the journey.

[27] Interview, Noel de Villiers.

[28] In 2005 it was involved in the development of routes in Barotseland (Zambia) and the Caprivi in Namibia.

[29] Open Africa does not sell the idea of routes to communities, but only gives support when invited to do so. This is in line with its principle that the routes should be owned by the community they serve.

[30] According to De Villiers, farmers were overproducing wine for an otherwise beer market and they needed to find innovative ways of attracting customers. They were not necessarily thinking in terms of wine routes *per se*.

[31] The route was jointly launched by Ministers Fernando Sumbana (Mozambique), Marthinus van Schalkwyk (South Africa) and Thandi Shongwe (Swaziland).

[32] allAfrica.com (accessed on 24/05/2006).

[33] RETOSA has three categories of membership, namely, full, associate and affiliate memberships. Public tourism authorities operating in Member States have a full membership.

[34] RETOSA is funded by the public sector, private sector and SADC Co-operating Partners.

[35] Blue, green, yellow and brown. The blue colour represents oceans, lakes, waterfalls and rivers; green for lush vegetation, forests and bush; yellow for the hot African sun, desert dunes and sand; while brown represents the African soil and cultural heritage (RETOSA 1999).

[36] It was hoped that the outcome of the conference would be to develop a 21st century African agenda for peace through tourism. That strategy should harness the immense power of tourism as a leading force for poverty reduction in Africa; identify needs to be met in order to implement socially, culturally, and environmentally responsible community tourism; showcase case studies of 'Success Stories' and models of best practice; and develop concrete action programmes to facilitate and nurture sustainable community tourism development (IIPT 2003b).

[37] Attending the Second IIPT Conference.

[38] The theme of the conference was 'Tourism: pathway to a peaceful and prosperous Africa'. As in the case of the IIPT Conference held in Tanzania, the IIPT Conference in Zambia aimed to develop a 21st century vision for African tourism and a strategy to achieve that vision.

[39] According to the IIPT (2005a), it was the world's first national peace through tourism week.

[40] The gap was a result of the unfinished railway project.

Chapter 8

Mapping Africa's future

Analysts of TFCAs should not only deal with a range of issues that arise from the creation of TFCAs, but should also engage with the conceptualisations of TFCAs. In other words, the analytical gaze should appreciate the different dimensions of TFCAs so as to enhance the theorisation of TFCAs and to develop appropriate and robust methods of evaluating them on the ground. At the level of theory, environmentalism and the concepts of peace, security and economic development that flow from it, provide a useful and multi-dimensional entry point to the study of TFCAs. The accentuation of the notion of ecological integrity in TFCAs implies that in theorising them we should consider debates in, and research results from, disciplines in the general category of natural science. The association of TFCAs with peace and security points to the need for TFCA studies to grasp theories of international relations while ecotourism in TFCAs calls for a deeper understanding of the workings of capital and its agencies in nature conservation. The main point here is that there is a need to engage with TFCAs at the abstract level and that the abstraction of TFCAs should be informed by, and reflect, the dimensions of TFCAs. Those dimensions range from TFCAs as imaginations through to their implementation and impact that they have on biodiversity protection, interstate relations, local communities, land use and ownership, and so forth.

TFCAs: an imagined geography

Hagen (2003) considers an imagined geography as ways of perceiving spaces and places, and the relationships between them, and as representing complex sets of cultural and political practices and ideas defined spatially. An imagined geography cannot be confined to static and discrete territorial units, all the more because spatially defined ideas and practices are continuously been shaped by a mixture of factors and actors operating at various geographical scales (Mudimbe 1988; Said 1994; Agnew 1999; Kitchin and Kneale 2001; Merrington 2001; Paasi 2002). However fictitious geographical imaginations might be they have a constitutive role to play in our everyday lives (Castree 2004). Arguably, the point of departure in understanding geopolitical intentions and interests is to examine closely their rootedness in geopolitical imaginations. The identification and decoding of a 'grand plan' is therefore a necessary precondition for establishing the links between imagined geographies and activities on the ground. Jhaveri (2004) made this point clear when arguing for a clear understanding of how oil enables the United States to pursue its petroimperialism. In a different but related context, Shalom (1999) used the foreign policy of the United States during the Cold War to suggest that a misunderstanding of the foundation of a political imagination could lead to the misinterpretation of the foreign policy of a particular country. He argued that, contrary to conventional wisdom, the Soviet Union was never the exclusive force behind US foreign policy. Rather, the policy arose from the imperial projections of the United States, which sought to construct and maintain a global stage to which the Soviet Union together with Third World nationalists and First World competitors were a threat. Indeed, the quest for control of the world stage accounts for US interventionism long before the collapse of the Soviet Union. This calls for the need to grasp the multiple sources that are mobilised in the construction of imagined geographies, and how these imaginations give content to processes and policies on the ground. TFCAs in Africa are an example of those geographies, hence they are correctly called an African Dream.

Africa's TFCAs do not only construct images of the continent for tourism purposes, but also seek to consolidate those images into a defined common feature to be pursued on the grounds of ecological integrity and as socio-economic and political imperatives. Images of Africa are not post-Cold War phenomena, as the colonisation of Africa was supported by an array of those images (Mudimbe 1988). Nonetheless, the images of the continent that emerged after the collapse of the apartheid state in South Africa and the end of the Cold War were part of a process that sought to shape the future of the continent under entirely new national, regional and global political environments. TFCAs emerged as part of the package that sought to determine how Africa should look. As the promotional magazine of the TFCA idea has admitted, the maps of TFCAs 'were not ordinary maps, but rather a vision: a window of how Africa might look like' (*Getaway* 2003, 62). That vision accounts for various nature-based activities on the continent, particularly in southern Africa. The TFCA imagery is embedded in the rearrangement of historical facts, socio-economic conditions and political movements according to a contemporary biodiversity grid. However, it would be naïve to explain the rise of TFCAs in southern Africa as a regional phenomenon. As I have argued in this volume, TFCAs can more appropriately be understood within the context of the environmental movement and the processes and practices associated with that movement. A closer inspection of that movement reveals that concerns with the ecological disintegration, and the part played by humans in that process, are central to environmentalism, as well as the approaches that various environmental groups have adopted.

Ecological integrity and TFCAs

The strongly held view that human survival and welfare depend on the fully functional ecological systems has inspired generations of environmentalists to find strategies for protecting and managing ecosystems, including attempts to recreate ecosystems where possible. The environmental movement has also paid, and continues to pay, a great deal of attention to the human activities that contribute to ecological disintegration. Industrialisation, population explosions, the development of nuclear weapons and the like, not only gave impetus to the Movement for Survival in the 1970s, but also led to the development of different approaches towards a blueprint for the future. Those approaches account for the emergence of different environmental interest groups. The groups include: NGOs, such as the IUCN, that seek to develop global ecological policies through governments while remaining outside the state system; Green parties that sought and still seek to take over the control of the modern state in order to establish a 'green state'; and a grassroots movement of bioregionalism that aims to radically reorganise the world by abolishing the state and its borders. There has also been a move towards developing international instruments for biodiversity protection such as the Convention on Biological Diversity so as to encourage, and even to force, governments to abide by international rules governing the use, protection and management of natural resources. To place TFCAs in Africa squarely into one of these approaches could be misleading, more especially because TFCAs are established and promoted through a careful selection of different threads of the environmental movement. For example, TFCAs are underpinned by the logic of ecological integrity that is central to the environmental movement, yet they are aimed at neither creating the 'green state' nor creating a world order based on bioregionalism. Clearly, TFCAs in Africa are not promoted through party political systems as there are no visible Green parties in the continent.

As elsewhere, TFCAs in Africa are part of the global solution to the environmental crisis. The *Africa Geographic* (2006) describes the establishment of TFCAs in Africa as the last attempt to link natural land for the protection of biodiversity patterns and processes. That aim is supportive of the idea of re-establishing ecoregions and of the whole process of bioregional planning, which has gained prominence through the analysis of cross-border environmental problems and the constraints that 'closed habitats' impose on species. From the ecological perspective, ecoregions do not only refer to large geographic units, but are also exclusively concerned with non-human species, habitats, environmental conditions and ecological dynamics. In contrast to the publicised definitions of TFCAs as areas combining protected areas and the cultural attributes, TFCAs in Africa are being promoted as ecoregions that are devoid of humans.[1] This is so, because there has been a shift from the concept of TFCAs to that of transfrontier parks. As I have shown in the various chapters of this volume, the establishment of TFCAs in the African continent, particularly in southern Africa, is based on the amalgamation of contiguous national parks. In the absence of national parks in the areas designated for TFCAs, national parks have been created as the core of the TFCAs, as the examples of the GLTP and the Limpopo/Shashe have shown. If indeed TFCAs are meant to preserve the cultural qualities of the areas in which they are established, there would have been no need for the resettlement of local residents. In contrast to the marketing of TFCAs as an initiative that respects and promotes the land rights of local residents, the residents are being resettled, or will face resettlement in future, precisely because the concept of TFCAs is increasingly being replaced by that of transfrontier parks, the implication being to create 'superparks'.

The obsession with ecoregions in TFCAs in southern Africa is also evident in the interests of proponents of TFCAs on the migratory routes of wildlife. Elephants are particularly used to define ecoregions for TFCAs. Accordingly, *Africa Geographic* (2006) has suggested that elephants define Africa's savanna and that the planet's wild places would be diminished without them. In this sense, elephants are instrumental in recreating the wilderness, which is central to the idea of TFCAs. In the process of creating the wilderness through elephants, human migration has been ignored in order to secure ecoregions. The result is to promote animal migration across political boundaries while, at the same time, persecuting humans who are found crossing the same boundaries. The most active participants in TFCAs, Botswana and South Africa, have strengthened their borders against immigrants from their neighbouring countries. Botswana has even built an electric fence along its border with Zimbabwe (Gumede 2005).

In the promotion of TFCAs the cost of these cross-border initiatives to humans is covered through the use of politically correct language – a rare occurrence in which the language of social science is used by natural scientists to promote a natural science agenda. In the analysis of TFCAs, the use of that language across disciplines has meant that there is collusion among scientists from different backgrounds and perspectives. Before we dwell on the discussion of the use of social science language in the promotion of TFCAs, some further comments on the establishment of TFCAs as ecoregions would be appropriate. Ecoregions are not simply spaces that are created for the protection and management of biodiversity, but embody the whole process of the production of space and the power dynamics that go with that process. This is clear from the discussion on the appropriate boundaries of ecoregions.

The number of ecoregions varies according to the factors that are used to define them, and also according to the purposes for which they are established. Apart from protecting viable ecosystems, Africa's ecoregions represent the process of creating space by grabbing land from a land-hungry constituent, with the promise that that process is in everybody's interest. Accordingly, the PPF regards the establishment of TFCAs as a process of securing space.[2] In practice, TFCAs secure space by denouncing state boundaries as artificial and human-made so as to draw new boundaries on ecological grounds. Read from this angle, ordinary nature conservation projects such as cross-border water management schemes have generally been ignored in the creation of TFCAs, because they do not assist in securing space with clear boundaries. More importantly, countries such as Kenya and Tanzania that do not have fences impeding the migration of wildlife have not been used to launch TFCAs, partly because their 'boundless' spaces would have to be defined first before bounded spaces could be created and secured from state control and intervention. Put differently, proponents of TFCAs claim that state boundaries should be removed in order to allow the re-establishment of the migratory routes of wildlife, yet, at the same time, they did not capitalise on the existing migratory routes that have not been obstructed by the so-called artificial boundaries. Arguably, unfenced areas do not lend themselves to the TFCA process, because that process creates fences through a process of de-legitimatising other fences.

The role of social science in the construction of TFCAs

As we have seen in Chapter 2, the environmental movement was strengthened by concerns over the negative effects of industrialisation on the earth's support systems. While such concerns were relevant to the industrialised countries of the North, the anti-industrialist stance had its limitations in most developing countries where the level of industrialisation is generally low. The strategy that the environmental movement adopted with regard to developing countries was to focus on, among other things, poverty as the main threat to ecological systems and human survival. By focusing on poverty as an enemy of the environment, the environmental movement has garnered strong support from individuals, groups and institutions that are concerned with poverty. Clearly, the links between poverty alleviation and nature conservation mean that TFCAs sit nicely in the global projects for poverty alleviation such as the Millennium Development Goals. Conceptually, the human dimension of the environmental crisis gained currency when the agendas of development and environment were pursued as two sides of the same coin in the late 1980s. Those agendas were sealed at world forums such the Rio Earth Summit (1992) and the Johannesburg Sustainable Development Summit (2002), where emphasis was placed on the mutual co-existence of environmental protection and development.

Ecotourism in TFCAs: a bioregional economy?

It follows that ecotourism in TFCAs represents the economic paradigm of the environmental movement. That paradigm has been made relevant to the African continent by adapting it to the socio-economic conditions in the continent. In environmental circles, poverty is viewed as one of the major threats to the environment. Flowing from this view is the use of ecotourism as the most appropriate method for poverty alleviation. Admittedly, ecotourism has the potential to contribute to the welfare of impoverished communities, but to suggest ecotourism as an absolute strategy for poverty alleviation and as the only viable mono-economy, particularly in rural areas adjacent to TFCAs, is to invalidate other forms of economic activities and livelihood strategies. Indeed, such a process of invalidation occurred in tandem with the promotion of TFCAs in southern Africa. For instance, various studies have been conducted in the region to emphasise the need to switch from agricultural farming to wildlife-based activities, such as game farming. Work has also been done to push the idea that there is nothing else for rural residents in or adjacent to areas designated for TFCAs to live on except ecotourism. There is nothing wrong with ecotourism *per se*, but the emphasis on ecotourism in TFCAs sites does not only fit into the form of economy that environmentalists generally support, but is also meant to extend capitalism into the remotest rural parts of, mostly, developing countries. More importantly, the emphasis on ecotourism in and around TFCAs sites also serves to acquire property, in the form of communal land, without the use of violence. That is to say, the violence through which land was acquired for protected areas has been replaced by less violent mechanisms that legitimise the predominantly Western property regime. Acquiring land for TFCAs through ecotourism is therefore a bloodless method, which Visvanathan (1998, 238) described in the following series of questions: 'why kill, when you can co-opt? Why destroy, when you can absorb the world through Keynesian strategies? Why police blatantly, when the expert advertisement can make the victim compliant?'

Interest in ecotourism in TFCAs has meant that the development of TFCAs and tourism routes has been realigned in order to make TFCAs tourist attractions. This is clear from the development of a network of tourism routes in the form of the Lubombo route that connects five TFCAs for tourism purposes. The route is promoted on the basis that the TFCAs and surrounding areas offer a diversified tourism product that would simultaneously benefit Mozambique, South Africa and Swaziland. As I have shown in Chapter 7, the Coastal Adventure Route links St Lucia Wetland Park to the Kruger National Park via southern Mozambique, the aim being to connect the Lubombo TFCAs with the GLTP. Moreover, the links between tourism and TFCAs in the region have an institutional support in the form of SADC and RETOSA. As a subsidiary organisation of SADC, RETOSA does not only encourage and facilitate the movement and flow of tourists into the region but is also responsible for branding the region. Its focus on scenic beauty, wildlife, culture and seaside means TFCAs are relevant to the activities of the region that RETOSA has branded.

Apart from harnessing Africa's rich natural and cultural heritage for the promotion of tourism, TFCAs are also linked to the concept of tourism for peace. Conceptually, the use of TFCAs as a catalyst for peace on the continent supports the dominance of the idea of a 'park' in the creation of TFCAs. In practice, TFCAs are marketed as peace parks for tourism purposes in line with the view that ecological integrity could be used to support peace between and among nations. Accordingly, the IIPT is at the forefront of establishing the links between peace parks and tourism in Africa by fostering the commitment by governments and the private sector to the social, cultural, environmental and economic benefits of tourism. In this sense the IIPT promotes the view that tourism is a catalyst for reconciliation and peace on the continent. To that end, it has endorsed the United Nations declaration of the first decade of the 21st century (i.e. 2001–2010) as the International Decade for a Culture of Peace and Non-Violence for the Children of the World. The importance of ecotourism in TFCAs goes beyond concerns with local economic development to the promotion of TFCAs as a moral and economic necessity at national, regional, continental and global levels. It should be emphasised that though the seed of TFCAs was planted in southern Africa, the roles that TFCAs play in tourism have been appreciated at the continental level as a key factor of the African economy. At the time of writing (May 2006), South Africa, as the host of the 2010 Soccer World Cup on behalf of Africa, was already planning to use tourism in TFCAs as one of the strategies that the rest of Africa could benefit from during the 2010 soccer event. The *Sunday Times* (2006a) has most recently published job advertisements for the Programme Manager,[3] Infrastructure Development Manager and Marketing Manager[4] for the newly created TFCAs 2010 Development Unit.[5] The impact of that event cannot be pre-empted, but there are indications that that event will be used to consolidate TFCAs as an engine of ecotourism in Africa. Future research on TFCAs and tourism in Africa will do well to establish the impact of the 2010 Soccer World Cup on the development of TFCAs and their resultant economic benefits.

In short, ecotourism in TFCAs conflates bioregionalism and capitalism. Bioregionalism, as we have noted in this book, is opposed to the capitalist economy and favours communal rather than private ownership of resources. The growing trend towards communal conservancies and ecotourism ventures, particularly in southern Africa, fosters the spirit of communal ownership; meanwhile communal land is incorporated into the global tourism industry.

Institutional transformation

The creation of TFCAs does not occur in an institutional vacuum, but the process penetrates through the state to the different actors operating at various, and interconnected, scales. The state provides the foundation for institutional transformation, including its own transformation in different ways. These include attempts to de-establish the state boundaries by rendering them useful for state sovereignty; meanwhile the natural resources that traditionally belonged to the state are placed in the hands of supranational entities and structures. The reasons advanced for doing so are that: the state is incapable of playing the role of the guardian of nature; environmental problems are not and cannot be limited to the boundaries of the state; political systems are based on short cycles; and the organisation of the state makes it impossible to co-ordinate efforts towards protecting and managing life-support ecological systems. Within environmentalism, the state is under attack from various quarters, ranging from the Green parties through peace movements to bioregionalists. Whereas Green parties seek to transform the state from inside, bioregionalists aim at the radical transformation of the state so as to re-establish the organisational structure of 'the native'. Such a radical agenda appeals to historically marginalised groups such as indigenous peoples as it reconfirms 'indigenous institutions' and histories. With regard to TFCAs, indigenous knowledge and institutions are not re-established for their own sake, but are mobilised to support the creation of supranational institutions and structures that govern natural resources. In the process of doing so, the histories of the migration of local residents are erased in favour of that of animals, and local communities are systematically marginalised, as I have argued in Chapter 6.

It could be argued that the process of institutional transformation is principally aimed at the state, though that process ironically operates through the state. As with the wider environmental movement, TFCAs are underpinned by the view that the rules of state sovereignty[6] are inadequate for biodiversity protection and management hence there is a need for a shared sovereignty of some kind. In his analysis of the institutional options for dealing with weak or failed sovereignty, Krasner (2004, 85) suggested that 'in future, better domestic governance in badly governed, failed, and occupied polities will require the transcendence of accepted rules, including the creation of shared sovereignty in specific areas'. Shared sovereignty could involve the engagement of external actors in some of the domestic authority structures of the target state for an indefinite period of time. Though the concept of shared sovereignty does not appear in treaties and MoUs governing TFCAs, its applicability to the governance and ownership of biodiversity cannot be dismissed, not least because TFCAs deterritorialise biodiversity.[7] In this regard Duffy (2006) correctly described TFCAs as a form of transnational management of the environment.

In the context of TFCAs, institutional transformation also implies changing the roles and powers of the state over natural resources, restructuring and streamlining government departments and ministries, harmonising legislative instruments among participating states, and reorganising and strengthening community governance structures. The creation of the Ministry for Co-ordination of Environmental Affairs (Mozambique), the Directorate for Transfrontier Conservation and Protected Areas (South Africa), and the transfer of the Department of Parks to the Ministry of Environment in Lesotho, are examples of the relevance of the restructuring of state institutions that facilitate the establishment of TFCAs. With regard to the legislative instruments necessary for TFCAs, different countries are involved in promulgating laws that make the creation of TFCAs possible. Most of the legislative instruments are in the areas of the environment and tourism. Against this backdrop, national policies that are promoted and implemented by states participating in TFCAs provide a useful background for analysing the links between state policy and TFCAs. As I have shown in Chapter 6, there is congruence between state policies and the envisaged map of TFCAs in southern Africa.

The cross-border nature of TFCAs implies that changes taking place within participating states should be directed towards achieving a supranational goal. That goal can only be realised through the establishment of institutions at the supranational level. In southern Africa, SADC provided the necessary conditions for the development of TFCAs for reasons that I have advanced in this book and that are summarised below. SADC provided member states with common political, economical and environmental agendas. The political agenda of forming a regional entity and identity meant that member states should act as members of the same community. As in most regional blocs, the economic rationales for SADC include the establishment of a common market and the development of rules and regulations governing trade relations among member states on the one hand and between SADC and non-members on the other hand. The SADC environmental agenda that translated into the SADC Protocol on Wildlife Conservation and Law Enforcement was particularly relevant for the establishment of TFCAs in the southern African region.

It could be argued that the political, economic and environmental concerns of SADC are not unique to the region, raising the question of why SADC has been a launching pad for TFCAs in the African continent. Admittedly, the process of regionalism in Africa is not confined to southern Africa, but has taken place in other parts of the continent as evidenced by regional blocs such as UMA, COMESA, ECOWAS and ECCAS, which are found elsewhere in the African continent. These regional blocs are founded on similar political and economic bases that led to, and that underpin, SADC. By the same logic, other parts of Africa also had environmental conditions that could have been used to support the ecological aims and objectives of TFCAs. Despite meeting all the requirements for TFCAs, those regional blocs have not been used in the promotion of TFCAs, a situation that has conferred leadership on southern Africa in TFCA initiatives in Africa. As I have argued in this book, SADC became a launching pad for TFCAs, mainly because the main force behind the establishment of TFCAs in Africa, the PPF, has a long history of nature conservation in southern Africa. SANF, as the frontrunner of the PPF, laid the foundation on which the PPF could launch TFCAs in southern Africa. In other words, the regional reach of SANF ensured that the PPF would inherit the necessary agencies, human and non-human capital and the infrastructure that could be mobilised for the creation of TFCAs. The use of these resources was geographically limited to southern Africa, particularly to South Africa and its immediate neighbouring countries. The PPF succeeded in promoting the concept of TFCAs in the region, because it has the support of SADC structures. The point here is that regionalism is not unique to southern Africa, but the combination of that process and the strategies and resources of the PPF has meant that southern Africa stands out as the region leading the development of TFCAs on the African continent. To be sure, regionalism in SADC provided the necessary conditions for the development of TFCAs in the southern part of the African continent. The transformation of SADC in the late 1990s in particular offered supporters of TFCAs the opportunities to sell their agenda to SADC structures. Current and future analyses of TFCAs in Africa are bound to confront the role of the PPF in the establishment of TFCAs.

Given the regional dynamics and rivalries between and among regional blocs in Africa, the launching of TFCAs in southern Africa is most likely to place SADC as a regional hegemony on TFCAs in the continent. Indeed, the leadership role that South Africa played in the creation of TFCAs has been interpreted as the manifestation and entrenchment of South Africa's dominant position in southern Africa. Equally, the view that TFCAs are being driven by whites in South Africa has the potential to generate the lack of interest in TFCAs in other parts of the continent. The point is that concerns with TFCAs as a South African or SADC initiative imply that there is a possibility that other regions might show a lack of interest in the TFCA initiative. That possibility is remote, because TFCAs have been adopted as part of the NEPAD Environment Initiative. The adoption of TFCAs and the recognition of TFCA agreements by NEPAD imply that African states participating in NEPAD have embraced the TFCAs' agenda. Against this backdrop, African states, donors and environmental NGOs will be able to promote TFCAs on the entire African continent as a NEPAD project. In this way, the future of TFCAs has been tied to that of NEPAD, and therefore to that of the continent as a whole.

Endnotes

1 The idea of the wilderness seems to re-emerging through the establishment of TFCAs.

2 See PPF website: www.peaceparks.org

3 The position involves providing leadership to and guidance to the TFCA 2010 Development Unit.

4 The Marketing Manager is expected to develop a marketable brand for the TFCA route, develop a marketing and implementation plan for the TFCA, identify potential tourism facilities in each TFCA, and liaise with various regional and national tourism authorities.

5 The Unit is attached to the South African Department of Environmental Affairs and Tourism.

6 A recognition of juridically independent territorial entities and non-intervention in the internal affairs of other states (Krasner 2004).

7 This takes place despite the emphasis by treaties and MoUs that the TFCAs respect the sovereignty of the state.

Chapter 9

Postscript: transfrontier parks and the legacy of Rupert

> The imprints of Anton Rupert's legacy are scattered all over southern Africa, further afield and across international boundaries. They cover the business world, cultural life and the natural environment. They are embedded deep in the hearts of all those who have crossed his path and have been enriched by the experience[1] (*Cape Times* 2006a, 9).

> Few societies in the world have had the privilege to produce individuals of the calibre and integrity of Dr Anton Rupert. He will be remembered, honoured and appreciated by generations to come for the profound legacy he left on a wide canvas in our own as well as international societies (*Business Report* 2006, 3).

The passing away of one of the strongest advocates for and financier of transfrontier parks in Africa, Anton Rupert, on 18 January 2006, did not only make headlines in the media in South Africa and beyond, but also brought the life and activities of Rupert to the attention of the general public. The Cape Town-based Afrikaans newspaper, *Die Burger*, ensured that the public was informed of Rupert's activities by providing free copies of the newspaper to the public,[2] with a supplement written in English. Media reports and comments on Rupert after his death are relevant to the theme of this book in two main ways. Firstly, they provide information on Rupert's involvement in national and transfrontier parks. That information is useful in closing or opening gaps in the analysis of TFCAs. Secondly, they provide public opinion on TFCAs and also shed light on the bases of that opinion. For these reasons, this postscript was considered a timely and necessary part of the book. On the whole the postscript is not intended to endorse or refute media reports. The aim is rather to tease out media reports and statements that enrich the analysis of this book and that suggest some of the research themes for future work on TFCAs. Specifically, the postscript pays attention to media reports on Rupert's vision and ambitions, his relationship with the apartheid state, his identity, his business activities, and his contribution to society and nature conservation. These were the themes that the media highlighted, and commented on.

Rupert's philosophy, vision and ambitions

Most media reports were based on the Rupert biography published in 2005. The main author of the biography and former editor-in-chief of *Die Burger*, Ebbe Dommisse, was one of the busiest persons shortly after the announcement of Rupert's death as he was considered the source of information about Rupert. The *Sunday Independent* and the *Star* published book reviews on Rupert's biography to commemorate his death. Dommisse used the first paragraphs in the first chapter of his book to suggest that Rupert's philosophy and vision was profoundly shaped by the horrors of World War II, particularly the dropping of the atomic bomb on Hiroshima in 1945.[3] The media emphasised the influence of the experiences of World War II on Rupert's life in the days following his death, calling it a Damascus experience (*Cape Times* 2006a). Rupert's philosophy of co-existence is ascribed to his reflection on the effects of that war and the impact that it would have on the future of the world. The *Sunday Times* (2006a, 1) summarised the impact of World War II on his philosophy and the relevance of that philosophy to peace parks in these words:

> 'Man [according to Rupert] had always dreamed about having power. Now he had got the power, but it was the power to destroy himself. I came to the conclusion that life would never be the same again; we would live like scorpions in a bottle.' Out of this realisation came his philosophy of co-existence and partnership, and his subsequent determination to create peace parks.

The atomic bomb is said to have changed Rupert's perception of life. However, his view on, and love for, nature developed from childhood with some religious connotations. Rupert developed his love for nature long before the outbreak of World War II. As a child, his father taught him that leaders of all monotheistic religions came from the desert,[4] meaning that the Karoo, Rupert's birthplace, has the potential to produce people who would 'naturally respect and appreciate nature' (*Cape Times* 2006b). Conceptually, this implies that Rupert's love for nature is neither a product of external influences nor part of his capitalistic endeavours. Rather his view on nature emerged from his conscience that developed in the Karoo in the Cape. This perspective does not only insulate Rupert from the general criticism of colonial and apartheid nature conservation philosophies and practices, but also elevates him above most white conservationists.[5] This is clear from Ian Player's comment that, 'he was one of the great Afrikaners of our time who really put an enormous amount of energy and money into culture and into conservation, and in that regard he stands out above all his peers' (cited in Yeld 2006, 31).

Out of his natural love for nature emerged ambition to leave this world a little better. In the light of this and the discussion in Chapter 2 of this book, Rupert can be seen as part of the peace movement that became stronger after World War II. That movement was fiercely opposed to war. It could be suggested that Rupert's notion of peace, which was later to be promoted through peace parks, had a foundation that is not only historical but that also goes beyond the imperatives of nature conservation. Of significance to the theme of this book is Rupert's philosophy of co-existence, which developed from his deep reflection on the consequences of the World War II to humans and non-humans.[6]

Apartheid, Rupert and the future of South Africa

In her review of Rupert's biography, Maureen Isaacson commented that the biography evaded the question of whether Rupert funded the National Party, which ruled South Africa under apartheid (*Sunday Independent* 2006a). That question was neither raised nor answered by the media at the time of Rupert's death. It appeared vaguely in a different column of the same newspaper, where Isaacson quoted the statement that Rupert made to the *Afrikaanse Handelsinstituut* in May 1977 in which he said: 'my organisation never donates any money to any political party in any country. I have followed the policy of never having any contact with anyone from my group if he becomes a minister of state' (*Sunday Independent* 2006a, 4). Moreover, the media also avoided asking the same question from another angle, i.e. whether the apartheid state funded Rupert's nature conservation projects and what the reasons for the funding might have been. Clearly, the 42 or so national parks that Rupert helped to establish were of interest to, and administered by, the apartheid state.

Interestingly, the media was preoccupied with proving that Rupert adopted an anti-apartheid stance. The *Star* (2006a, 1) wrote that 'Rupert was well known for his opposition to apartheid's inhumanities', while the *Sunday Independent* (2006b) reported that Rupert clashed with the fiercest of apartheid rulers, Hendrik Verwoerd and Pik Botha. The *Star* (2006b) presented a clear picture of Rupert's anti-apartheid stance in these words:

> He played an important role in trying to convince leaders like John Vorster and PW Botha that policies should be changed … 'reaffirm your rejection of apartheid [Rupert said]. It is crucifying us; it is destroying our language; it is degrading a once heroic nation to be the lepers of the world. Remove the burden of the curse of a transgression against mankind from the backs of our children and their children … [and he wrote to Botha after his Rubicon speech of 1986] should you fail in this God-given task, then one day we shall surely end up with a Nuremberg.'

Given that enemies of the apartheid state of all backgrounds were harassed, hunted down, maimed and killed, the media was at pains to explain why the agencies of the apartheid state did not harm Rupert as it did others, including anti-apartheid activists from the Afrikaner community. The explanation offered is that Rupert did not opposed apartheid loudly, because he wanted to protect his business interests (*Die Burger* 2006).

There is an acknowledgement that Rupert was once part of the Afrikaner secret organisation and think-tank of the apartheid state, the *Broederbond*, and that he later abandoned the organisation,[7] calling it 'an absurdity' (*Sunday Times* 2006b, 15). That acknowledgement was not accompanied by the reasons why he joined the organisation in the first place, because to do so might expose cracks in his philosophy and practices. The emphasis on his withdrawal from the *Broederbond* does not only suggest his anti-apartheid stance, but also makes him a candidate of non-racialism in South Africa. Accordingly, the media emphasised the view that Rupert believed in a non-racial future for South Africa, and that had he assumed the leadership of the apartheid state, South Africa would have followed a different path of history (*Cape Times* 2006a). The *City Press Business* (2006, 3) emphasised that point by commenting that 'if Anton became prime minister[8] in Hendrik Verwoerd's place, [South Africa] would have been totally different'. He would have been too sensitive to implement apartheid's brutal laws and policies. The implication is that he was an antithesis of apartheid leaders. He was, as the Democratic Party leader, Tony Leon, described him, 'the ultimate renaissance man' (*Cape Argus* 2006, 1) who carried the hope of South Africa. The *City Press Business* (2006, 3) went to the extent of saying that, 'more than one person – from former President Nelson Mandela to futurologist Philip Spies – has observed that as long as [South Africa] has people of Anton Rupert's calibre, it would remain a country of hope'.

The 'non-materialist capitalist'

It is said that World War II made Rupert realise two important things, the first being that war creates the market for liquor and tobacco. His business empire emerged as a response to that market. Accordingly, the *Cape Times* (2006a, 9) reported that, 'having experienced the 1930s depression and having seen the effects of World War 2, [Rupert] realised that these commodities [i.e. liquor and tobacco] sold in larger quantities than lesser quantities during hard times'. The point here is that Rupert's business empire was founded on his profound understanding of the market. 'He had', as the *Sunday Times* (2006a, 15) put it, 'a genius for marketing and he never missed a trick'. According to a statement released by his family, 'he knew the value of international renown trademarks long before concepts such as "intellectual property" and "intangible assets" became fashionable' (*Star* 2006a, 1).

Unlike all commentators in the media who lauded Rupert's entrepreneurship, which he developed from 'his humble beginning', Richard Okes and W.L. Mason published letters in the *Mail & Guardian* newspaper, condemning the health risks that Rupert's tobacco and liquor pose. Okes (*Mail & Guardian* 2006, 22) condemned news editors who praised Rupert's business in these words: 'don't editors hear graveyards groan with rotten lungs and livers, retirement villages wheeze, cancer wards croak? Can't they feel the despair of countless families as they reel from alcohol abuse – broken-spirited men, beaten women, molested and neglected children?' Mason added to the criticism by saying that 'Rupert might have been a bastion of conservation and of the arts, but to me anyone who waxes rich from peddling a proven killer – in this case cigarettes – is an unashamed opportunist' (*Mail & Guardian* 2006, 22). Negative comments such as these were an exception in the media. Almost all media reports praised him for doing business with a sense of morality. Rupert's son, Johann Rupert, is reported to have said that his father believed in the free-market system and in sharing wealth generated from that system (*Star* 2006c). The following comments are worth noting:

> [Rupert] loved South Africa and was deeply committed to the empowerment of its people (Cyril Ramaphosa, cited in *Cape Times* 2006b, 1).
>
> He believed firmly in affirmative action and black economic empowerment. He told me once that whites would benefit from society when the cushion of public service as (the) first and frequently (the) only opportunity was denied to them (Trevor Manual, cited in *Star* 2006a, 1).
>
> Rupert had not only distinguished himself in the Afrikaner community, but also played a significant role in supporting and initiating the transformation of South African business (Thabo Mbeki, cited in *Cape Argus* 2006, 1).

The quotations given above were used by the media to illustrate Rupert's commitment and relevance to the now fashionable concepts of black economic empowerment and transformation in the business sector. His practical steps towards those political and commercial ideals were demonstrated by the examples that he once established a partnership company with the coloured community (*Sunday Independent* 2006b), set up a network of Indian retailers for his snuff through contacts with the anti-apartheid activist Yusuf Cachalia (*Sunday Times* 2006a), believed that blacks should be allowed to farm and own land, donated

R20,000 to victims of the urban violence in the 1970s (*Sunday Independent* 2006b), promoted equal rights during apartheid (*Guardian* 2006), and his Rembrandt company introduced 'a minimum wage of R2 a day in 1963, despite opposition from other employers' (*Sunday Independent* 2006b, 4). The media demonstrated his sympathy to blacks by highlighting that he initiated and funded language courses on records in 1964 'to encourage the learning of Xhosa, Zulu and Northern Sotho' (*Cape Times* 2006a, 9). His understanding and support for black economic empowerment stems from his experience of the marginalisation of the Afrikaner in the English-dominated market in the past (*Sunday Independent* 2006b).

According to media reports, Rupert saw the security and future of whites in South Africa as dependent on that of black people. He is reported to have said that if black people have nothing to eat then whites would not be able to sleep (*Cape Times* 2006a). The Premier of the Western Cape, Ebrahim Rasool, reiterated that statement during his opening address to the Provincial Legislature on 19 January 2006 not only to pay tribute to Rupert, but also to suggest that Rupert's vision for the indivisibility of South Africans should spur the white community to contribute towards building a shared future of South Africa that Rupert cherished.

The Afrikaner, the white South African, the African legend

> Born to sparkle as an Afrikaner nationalist, serving an exclusive cause, Anton Rupert was sufficiently inspired by the spirit of righteousness and justice, which is an inherent part of the making of the Afrikaners, to end his life as a celebrated South African and African patriot, a prophet of an inclusive future for the children of our country and continent (Thabo Mbeki 2006, 3).

In a country such as South Africa where contested identities are continuously created and recreated, one's position in society is often subjected to identity politics. A closer reading of media reports about Rupert's identity reveals that he has been made to simultaneously represent Afrikanerdom and the new South African identity that has become a national project in the democratic dispensation. His identity as an Afrikaner came out clearly from comments from the student leadership of the University of Pretoria, Rupert's alma mater. According to the President of the Student Representative Council at that university, Jan Hendrik Kay, Rupert was not only a pioneer in the Afrikaner community but also contributed to the rich traditions of that university, including the ox wagon,[9] conserving culture and cultural treasures – especially the presidential sash belonging to President Kruger, which hung in the offices of the SRC at the University of Pretoria – and the reintroduction of Afrikaans at that university. He was and remained proud of his Afrikaner-ness and the Afrikaans language (*City Press Business* 2006, 3). From the ideological point of view, it is said that Rupert started his business in partnership with Afrikaner nationalists such as Nico Diederichs in order to help the *volk* to escape from poverty and to challenge the hegemony of the English in business in South Africa (*Sunday Times* 2006b). In other words, Afrikanerdom was key to his impetus (*Sunday Independent* 2006b).

Rupert has also been described as a person whose national identity transcended boundary identities at different scales, and that his identity is not imposed, because he saw himself as simultaneously an Afrikaner, South African, African, world citizen and Christian[10] (*City Press Business* 2006). Moreover, his identity as an Afrikaner had nothing to do with white superiority. Rather, it embraces his vision for a non-racial South Africa (*Cape Times* 2006b), which is encapsulated in his belief that people are like grains of sand that are no better than the other (*City Press Business* 2006). As a result of his support for non-racialism and his willingness to work with black governments in southern Africa, the former *Broederbond* chairman and South African Broadcasting Corporation boss, Piet Meyer, described Rupert as a 'black brother' (*Sunday Times* 2006a).

Describing Rupert as part of the rainbow nation helps to claim his achievements as that of the South African nation. That is to say that he did not only put South Africa on the international stage, but he also proved that South Africans can distinguish themselves and successfully compete globally (*Cape Times* 2006b). He identified with the African continent and wished South Africa to contribute towards stability in the continent. He 'saw himself as a white African with a responsibility to the [African] continent' (*Sunday Independent* 2006b, 4). As a global citizen, he served the world and humanity.

The description of Rupert's political identity above has implication for analyses that seek to challenge TFCAs as some kind of an Afrikaner conspiracy. Such analyses will have to demonstrate that Akrikaners such as Rupert did not believe in the democratic dispensation. The challenge for critical research on TFCAs is to develop a credible historical analysis of the connection between TFCAs and the ideas and aspirations that emerged at particular moments. To criticise TFCAs in southern Africa and the continent at large on shaky grounds is most likely to render much of the work on TFCAs irrelevant and misinformed.

Peace parks: a monument

South Africa's Environment and Tourism Minister, Marthinus van Schalkwyk, commented that the passing away of Rupert marked an end to the era of global conservation (*Cape Argus* 2006), the point being that Rupert played a critical role in shaping the agendas for nature conservation beyond the shores of South Africa. Almost all media reports contained statements about Rupert's involvement in transfrontier parks, and presented these parks as the most important achievement in his life. The following statements are worth noting:

> I think he really was a builder and architect of the South African economy, of *wildlife* and the arts. He had a deep understanding of the people and soil of this land (Tony Leon, cited in *Cape Argus* 2006, 1).
>
> Not only will he be remembered for his business acumen, but also for his total devotion to nature and environmental conservation as shown by his immense contribution to the establishment of transfrontier parks. A true philanthropist (Thabo Mbeki, cited in *Cape Argus* 2006, 1).
>
> One of his most visionary projects was the peace parks, which will stand as one of his greatest monuments (Nelson Mandela, cited in *Cape Times* 2006c, 1).
>
> There could be no argument against Rupert's unrivalled contribution to conservation. He brought South Africa under the World Wide Fund for Nature (WWF) and is responsible for the creation of cross-border peace parks by collapsing borders between some of southern Africa's largest national parks. He pronounced this the most gratifying part of his life (*Sunday Independent* 2006b, 4).

There are salient points that the media raised with regard to Rupert's involvement in transfrontier parks, the first being that the idea of transfrontier parks had its origin in Rupert's creative thinking.[11] In that way he is the founder of the African dream. As the Chief Executive of the PPF, Willem van Riet, puts it, 'he was the man who produced the dream and he asked us to turn it into a reality, and that partnership between the dreamers and those making it a reality was extremely strong' (*Cape Argus* 2006, 2). In an apparent attempt to enhance the political credibility of the idea of transfrontier parks, the *Star* (2006a, 1) went to the extent of suggesting that Rupert cofounded the Peace Parks Foundation with Mandela and the late Prince Bernard of the Netherlands.

The second point is that transfrontier parks are founded on Rupert's philosophy of co-existence: the partnership between 'man' and 'man' on the one hand, and 'man' with nature on the other hand[12] (*Cape Times* 2006a). Nevertheless, what is meant by co-existence in transfrontier parks is vague, and lends itself to different interpretations. While it appears to challenge the division between nature and culture by suggesting the 'man–nature' interrelationship, the experiences of forced removals from, and the obsession with the recreation of the wilderness through, transfrontier parks in southern Africa render the notion of co-existence problematic.

The third point is that the first move towards realising one of Rupert's dreams for transfrontier parks was the creation of the GLTP (*Cape Times* 2006a). This point has been reiterated in Yeld's (2006, 31) comment that Rupert developed the vision of a series of peace parks 'when he saw the opportunities offered by a peace park encompassing the Kruger National Park and massive natural areas in adjoining Mozambique and Swaziland'.

Fourthly, Rupert was motivated to create transfrontier parks in order to create jobs for local communities through tourism. Media reports suggest that Rupert has always been concerned with poverty alleviation, and that his aim is mainly to help people to take the future into their own hands. For example, he was shocked by the extent of poverty among the Afrikaners as reported in the Carnegie Report (*Sunday Times* 2006b) and 'endeavoured to alleviate the appalling living conditions in the townships' through the Urban Foundation, which he co-founded (*Business Report* 2006, 3). Job creation through transfrontier parks is therefore a continuation of Rupert's philosophy of small business as 'a source of wealth creation, a form of empowerment that involved more than just the transfer of capital goods and share options that have become so prevalent in South Africa' (*City Press Business* 2006, 3). The stories of his contribution towards the upliftment of poor people, including blacks, have some relevance to the current marketing of transfrontier parks as a strategy for poverty alleviation. In other words, the idea that transfrontier parks would contribute to socio-economic development in impoverished black rural areas is presented as part of Rupert's nationally and globally tested business philosophy.

Fifthly, there is a strong message that transfrontier parks cannot be wrong, mainly because they are founded by one of the most successful people in the country and the world, who is well known for taking right and principled decisions. That is to say that private sector investment in TFCAs is a wise decision, because it is led by one of the pioneers 'in the global financial and commercial world' (*Cape Times* 2006b, 1), who has 'a unique sense of knowing what the consumer needed and that product quality was non-negotiable' (*Cape Times* 2006a, 9). The implication for TFCAs is that they are a good commercial product for those who want to invest in them.

Sixthly, the statement that Rupert was 'a man who tackled the challenges of the future with wisdom and experience of the past' (*Cape Argus* 2006, 1) implies that transfrontier parks are a blueprint for the future of the continent. At the level of abstraction, those who challenge transfrontier parks as a way to the future will have to provide an alternative and bright future for the continent. President Mbeki's description of Rupert as a 'prophet' also suggests that the continent should follow on the blueprint of transfrontier parks that Rupert has drawn for the continent. In bidding farewell to Rupert, Mbeki (2006, 3) wrote 'I trust that all of us, black and white, will find it within our capabilities to translate [Rupert's] prophecy into the lived and living experience of all the children of Africa'. Against this backdrop, transfrontier parks are here to stay as the lasting legacy of Rupert's achievements and vision for the continent (*Star* 2006a).

Lastly, Rupert has been elevated to the status of a hero (*Sunday Independent* 2006b) and words such as 'colossus' have been used to describe him as an extraordinary person. In paying tribute to Rupert, the *Africa Wildlife* magazine suggested that 'words like "colossus" or "legend" should never be used lightly. But in the case of Anton Rupert, the Afrikaner business tycoon and philanthropist extraordinaire … they are entirely apposite' (Yeld 2006, 31). Descriptions such as 'the colossus', which have been used for people such as Cecil Rhodes, imply that Rupert and TFCAs in Africa are inseparable, with Rupert being the hero of Africa. There is a possibility that the association between the Cape to Cairo concept and TFCAs could be used to imply that, while Rhodes dreamed of linking Africa under colonialism, Rupert has achieved the same goal in a liberated Africa. It is in the Cape to Cairo concept that the comparison between Rhodes and Rupert might in future be drawn.

Endnotes

1 Media statement released by Rupert's family.

2 The author received his free copy at the Pick 'n Pay supermarket.

3 It is said that Rupert was so saddened by the bomb that, for the first time, he stayed away from work for the whole day.

4 Such as Christ, Moses and Muhammad (*Cape Times* 2006a).

5 Most of them had to be trained to look after nature, and misused their power in their service to nature conservation agencies and organisations.

6 More comments on co-existence and transfrontier parks are provided in the sections below.

7 By not renewing his membership.

8 He had all the qualities to become one, and the eminent Afrikaans poet, Van Wyk Louw, once requested him to consider standing as a candidate for the Premiership (*Cape Times* 2006a).

9 He made it possible that the ox wagon that belonged to the Voortrekkers' leader, Louis Trichardt, could be preserved at the University of Pretoria. The ox wagon had been kept at the town of Louis Trichardt, which has since been renamed as Makhado. In so doing, he protected the ox wagon from being lost to the Afrikaner community, especially in a democratic South Africa where symbols of the apartheid past are being challenged by the previously oppressed majority, including the temptation to destroy them completely. The ox wagon is an important symbol of Afrikaner identity. It signifies the sacrifices of the previous generation of Afrikaners (*Die Burger* 2006).

10 It is said that Rupert did not see contradictions in these identities (*City Press Business* 2006).

11 He is, after all, a creator of ideas (*Star* 2006c).

12 Without this partnership, humanity will not survive.

References

Aberley, D. (1999) Interpreting bioregionalism: a story from many voices. In: Mcginnis, M.V. (ed.) *Bioregionalism*. Routledge, London, UK, pp. 13–42.

Addison, G. (1997) A dream without frontiers. *Leadership* 16, 65–74.

Adepoju, A. (2001) Regional organisation and intra-regional migration in Sub-Saharan Africa: challenges and prospects. *International Migration* 39, 45–59.

Africa Geographic (2005) Peace Parks Foundation Review (July), 52–53.

Africa Geographic (2006) Special Report on elephants & us (April).

African Security Review (1998) Editorial. *African Security Review* 7, 2.

African Union (2002) Protocol Relating to the Establishment of the Peace and Security Council of the African Union. 9 July, Durban, South Africa.

Agnew, J. (1999) Regions on the mind do not equal regions of the mind. *Progress in Human Geography* 23, 91–96.

Ali, A. (2002) A Siachen Peace Park: the solution to a half-century of international conflict? *Mountain Research and Development* 22, 316–319.

Ali, S.A. (Forthcoming) Introduction: a natural connection between ecology and peace? In: Ali, S.A. (ed.) *Peace Parks: Conservation and Conflict Resolution*. MIT Press, Cambridge MA, USA.

Anderson, J. and O'Dowd, L. (1999) Borders, border regions and territoriality: contradictory meanings, changing significance. *Regional Studies* 33, 593–604.

Andreas, P. (2000) *Border Games: Policing the US–Mexico Divide*. Cornell University Press, Ithaca NY, USA.

Asiwaju, A.I. (1985) Partitioned culture areas: a checklist. In: Asiwaju, A.I. (ed.) *Partitioned Africans: Ethnic Relations Across Africa's International Boundaries 1884–1894*. Hirst and Co, London, UK, pp. 252–259.

Bailey, R.G. (2005) Identifying ecoregion boundaries. *Environmental Management* 34, 514–526.

Baker, H. (1934) *Cecil Rhodes*. Oxford University Press, London, UK.

Baldus, R., Kibonde, B. and Siege, L. (2003) Seeking conservation partnerships in the Selous Game Reserve, Tanzania. *Parks* 13, 50–61.

Baldus, R.D. and Hahn, R. (Forthcoming) Connecting the world's largest elephant ranges: the Selous-Niassa Corridor. In: Ali, S.A. (ed.) *Peace Parks: Conservation and Conflict Resolution*. MIT Press, Cambridge MA, USA.

Banfield, J. (1999) Naidoo message spans Africa. *African Business* (June), 24–26.

Bang, M. (2002) Permaculture design: philosophy and practice. In: Jackson, H. and Svensson, K. (eds) *Ecovillage Living: Restoring the Earth and her People*. Green Books, Dartington, UK, pp. 18–19.

Barnard, P. (1998) *Biological Biodiversity in Namibia: A Country Study*. Namibian National Biodiversity Task Team, Windhoek, Namibia.

Barnes, J.I. and De Jager, J.L.V. (1995) *Economic and Financial Incentives for Wildlife Use on Private Land in Namibia and the Implications for Policy, Research Paper 8*. Ministry of Environment and Tourism, Windhoek, Namibia.

Beauregard, R.A. (1995) Theorising the global-local connection. In: Knox, P. and Taylor, P.J. (eds) *World Cities in a World System*. Cambridge University Press, Cambridge, UK, pp. 232–248.

Belcher, M.L. (1932) *Cape to Cowley via Cairo in a Light Car*. Methuen and Co, London, UK.

Bello, A. (1995) The boundaries must change. *West Africa* (10–16 April), 546.

Bennett, A.F. (2003) *Linkages in the Landscape: The Role of Corridors and Connectivity in Wildlife Conservation*. IUCN, Gland, Switzerland.

Besançon, C. and Hammill, A. (Forthcoming) Measuring peace performance: definitions and experiences. In: Ali, S.A. (ed.) *Peace Parks: Conservation and Conflict Resolution*. MIT Press, Cambridge MA, USA.

Blackie, R. (1999) *Communities and Natural Resources: Trends in Equitable and Efficient Use. Research Discussion Paper 29*. Ministry of Environment and Tourism, Windhoek, Namibia.

Botswana (1968) *National Development Plan*. Government Printer, Gaborone, Botswana.

Botswana (1985) *National Development Plan*. Government Printer, Gaborone, Botswana.

Botswana (1997) *National Development Plan 8, 1997/98–2002/3*. Gaborone, Botswana.

Botswana (1998) *National Settlement Policy*. Government Printer, Gaborone, Botswana.

Botswana (2001) *National Atlas*. Department of Surveys and Mapping, Gaborone, Botswana.

Bowen, W.M. and Wells, M.V. (2002) The politics and reality of environmental justice: a history and considerations for public administrators and policy makers. *Public Administration Review* 62, 688–698.

Braack, L. (2006) Conservation International's mega-elephant project: creating corridors across boundaries. *Africa Geographic* (April), 9.

Brenner, N. (2001) The limits to scale? Methodological reflections on scalar structuration. *Progress in Human Geography* 25, 591–614.

Breytenbach, W. (1999) The history and destiny of national minorities in the African renaissance: the case for better boundaries. In: Makgoba, W.M. (ed.) *African Renaissance*. Mafube, Sandton, South Africa.

Bright, C. (1911) *Imperial Telegraphic Communication*. P.S. King and Son, London, UK.

Broad, R. (1994) The poor and the environment: friends or foes? *World Development* 22, 811–822.

Brock, L. (1991) Peace through parks: the environment on the peace research agenda. *Journal of Peace Research* 28, 407–423.

Brockington, D. (2006) The politics and ethnography of environmentalism in Tanzania. *African Affairs* 105, 97–116.

Brooks, T., Balmford, A., Burgess, N., Fjeldsa, J., Hansen, L.A., Moore, J., Rahbek, C. and Williams, P. (2001) Toward a blueprint for conservation in Africa. *Bioscience* 51, 613–624.

Brown, J.C. and Purcell, M. (2005) There's nothing inherent about scale: political ecology, the local trap and the politics of development in the Brazilian Amazon. *Geoforum* 36, 607–624.

Brubaker, E. (1995) *Property Rights in the Defence of Nature*. Earthscan, London, UK.

Brunckhorst, D.J. (2002) *Bioregional Planning*. Routledge, London, UK.

Buell, F. (2003) *From Apocalypse to Way of Life: Environmental Crisis in the American Century*. Routledge, New York, USA.

Bullard, R.D. (1990) *Dumping in Dixie: Race, Class and Environmental Quality*. Westview Press, Boulder CO, USA.

Burgess, N., Hales, J.D., Underworld, E., Dinerstein, E., Oslon, D., Itoua, I., Schipper, J., Ricketts, T. and Newman, K. (2004) *Terrestrial Ecoregions of Africa and Madagascar: A Conservation Assessment*. Island Press, Washington DC, USA.

Business Report (2006) Business pioneer's career spanned old and new SA – and the world (20 January). Johannesburg, South Africa.

Butts, K.H. (1994) *Environmental Security: A DOD Partnership for Peace*. Strategic Studies Institute (US Army College), Carlisle PA, USA.

Cape Argus (1962) Monument to the 'Colossus' is part of the South African heritage (21 June). Cape Town, South Africa.

Cape Argus (2006) Anton Rupert: a giant passes (19 January). Cape Town, South Africa.

Cape Times (2006a) Anton Rupert: a man of worth (20 January). Cape Town, South Africa.

Cape Times (2006b) Mooi loop Anton, says Mandela as tributes pour in (20 January). Cape Town, South Africa.

Cape Times (2006c) Hundreds at funeral of Anton Rupert (26 January). Cape Town, South Africa.

Carr, M. (2004) *Bioregionalism and Civil Society: Democratic Challenges to Corporate Globalism*. UBC Press, Vancouver, Canada.

Carruthers, J. (1995) *The Kruger National Park: A Social and Political History*. University of Natal Press, Pietermaritzburg, South Africa.

Carter, A. (1992) *Peace Movements: International Protest and World Politics since 1945*. Longman, Harlow, UK.

Castree, N. (2001) Socialising nature: theory, practice and politics. In: Castree, N. and Braun, B. (eds) *Social Nature: Theory, Practice and Politics*. Blackwell, Oxford, UK, pp. 1–21.

Castree, N. (2004) Differential geographies: place, indigenous rights and 'local resources'. *Political Geography* 23, 133–167.

Castree, N. (2005) The epistemology of particulars: human geography, case studies and 'context'. *Geoforum* 36, 541–544.

Cawthra, G. (1997) Towards [a] holistic approach to security management at regional security level. *African Security Review* 6, 58–61.

CESVI (2003) *Sustainable Development and Natural Resources Management in Southern Zimbabwe*. CESVI Office, Harare, Zimbabwe.

CESVI (2005) *Sustainable Development and Natural Resource Management in Southern Zimbabwe*. CESVI Office, Harare, Zimbabwe.

Chapin, M. (2004) A challenge to conservationists. *World Watch* (December), 17–31.

Cheru, F. (2002) *African Renaissance: Roadmaps to the Challenges of Globalization*. David Philip, Cape Town, South Africa.

Chissano, J. (1998) Speech at the launch of Lubombo Spatial Development Initiative (6 May). Durban, South Africa.

Chris Mulder Associates. (1986) Master Plan and Management Guidelines for the Songimvelo Natural Resource Area. KaNgwane, South Africa.

Christopher, A.J. (1994) *The Atlas of Apartheid*. Routledge, London, UK.

City Press (2005) Hundreds flood through new border gate (18 December), 6. Johannesburg, South Africa.

City Press Business (2006) In Anton Rupert, South Africa has lost an indispensable man (22 January). Johannesburg, South Africa.

Clapman, C. (1996) Boundary and territory in the Horn of Africa. In: Nugent, P. and Asiwaju, A.I. (eds) *African Boundaries: Barriers, Conduits and Opportunities*. Pinter, London, UK, pp. 237–250.

Cleaver, T. (1997) *Understanding the World Economy: Global Issues Shaping the Future*. Routledge, London, UK.

Coates, P. (1998) *Nature*. Blackwell, Oxford, UK.

Coleman, J.S. (1994) *Nationalism and Development in Africa*. University of California Press, Berkeley CA, USA.

Colonial Office (1884) *Minute No.3 (July), CO 4532*. London, UK.

Conca, K. (1995) Greening the United Nations: environmental organisations and the UN system. *Third World Quarterly* 16, 441–457.

Copeland, M. (2003) *Cape to Cairo*. Sunbird, Llandudno, South Africa.

Coulter, G. and Mubamba, R. (1993) Conservation of Lake Tanganyika, with special reference to underwater parks. *Conservation Biology* 7, 678–685.

Dalby, S. (1992) Ecopolitical discourse: 'environmental security' and political geography. *Progress in Human Geography* 16, 503–522.

D'Amore, L. (2003) Welcome Address to the Second IIPT African Conference (8 December). Dar es Salaam, Tanzania.

Dauncey, G. (1988) *Beyond the Crash: The Emerging Rainbow Economy*. Greenprint, London, UK.

DeLuca, K.M. (2001) Rethinking critical theory: instrumental reason, judgement, and the environmental crisis. *Environmental Ethics* 23, 307–326.

De Villiers, B. (1999) *Peace Parks: The Way Ahead*. Human Sciences Research Council, Pretoria, South Africa.

De Villiers, D. (2003) Community tourism: the way ahead for Africa. Address to the Second IIPT African Conference (8 December). Dar es Salaam, Tanzania.

De Villiers, N. (1995) *The African Dream Project Handbook: Unleashing the Spirit of Africa*. Open Africa, Cape Town, South Africa.

Diarrah, A.S. (2002) The concept of the 'cross-border area' in the West African sub-regional integration process: results of the Sikasso Seminar (20–21 May), Accra, Ghana.

Die Burger (2006) SA groet sy 'eerste ware, groot entrepreneur' (20 January). Cape Town, South Africa.

Discovery SA (2006) In the saddle at Songimvelo (Summer/Spring), 36 and 40.

Dodds, K. (2001) Political geography III: critical geopolitics after ten years. *Progress in Human Geography* 25, 469–484.

Dommisse, E. and Esterhuyse, W. (2005) *Anton Rupert: A Biography*. Tafelberg, Cape Town, South Africa.

Doolan, J. (2006) Anton Rupert. *Leadership* (December), 18–22.

Downes, D. (2000) The New Zealand Enviromental Movement and the politics of Inclusion. *Australian Journal of Political Science* 35, 471–491.

Draper, M., Spierenburg, M. and Wells, H. (2004) African dreams of cohesion: elite pacting and community development in transfrontier conservation areas in southern Africa. *Culture and Organisation* 10, 341–353.

Drive Out (2005) What's up in the 4x4 world? (April–June), 8.

Dryzek, J.S., Downes, D., Hunold, C., Schlosberg, D. and Hernes, H. (2003) *Green States and Social Movements: Environmentalism in the United States, United Kingdom, Germany & Norway*. Oxford University Press, Oxford, UK.

Dube, M. (2001) Balancing sustainable ecotourism – an interview with the Minister of Environmental Affairs and Tourism, Valli Moosa. *Enterprise* (September), 26–28.

Duffy, R. (2001) Peace parks: the paradox of globalisation. *Geopolitics* 6, 1–26.

Duffy, R. (2006) The potential and pitfalls of global environmental governance: the politics of transfrontier conservation areas in southern Africa. *Political Geography* 25, 89–112.

Duncan, D.E. (1989) *From Cape to Cairo*. Weidenfeld and Nicholson, New York, USA.

Dzingirai, V. (2004) *Disenfranchisement at Large: Transfrontier Zones, Conservation and Local Livelihoods*. IUCN–ROSA, Harare, Zimbabwe.

Earthyear (2005) Tree Power. *Earthyear* l, 12–14.

East African Community Secretariat. (2001) The East African Community Development Strategy 2001–2005. Arusha, Tanzania.

Ecologist (1972) A blueprint for survival (special issue).

Economic Community of West African States (ECOWAS) (1975) *Treaty of ECOWAS*. Lagos, Nigeria.

Edwart-Smith, C. (2005) Greater Limpopo Transfrontier Park: hope after all the hype. *Drive Out* 10, 25–33.

Ellis, S. (1994) Of elephants and men: politics and nature conservation in South Africa. *Journal of Southern African Studies* 20, 53–69.

Emel, J. (2002) An inquiry into the green disciplining of capital. *Environment and Planning A* 34, 827–843.

EUROPARC (2003) *Newsletter No. 5*. Grafenau, Germany.

Fabricius, P. (2006) South Africa should back moves to report Iran to UN, says US. *Cape Times* (26 January), Cape Town, South Africa.

Fairhead, J. and Leach, M. (1996) *Misreading the African Landscape*. Cambridge University Press, Cambridge MA, USA.

Fall, J. (2005) *Drawing the Line: Nature, Hybridity and Politics in Transboundary Spaces*. Ashgate, Aldershot, UK.

Ferrão, G. (2003) *The Great Limpopo Transfrontier Park Project Report: July–December 2003*. Maputo, Mozambique.

Ferreira, S.L.A. (Forthcoming) Communities and transfrontier parks in the Southern African Development Community: the case of the Limpopo National Park, Mozambique. *South African Geographical Journal.*

Finland (2004) *Statistical Yearbook of Finland*. Helsinki, Finland.

FitzRoy, F. and Smith, I. (2004) Welfare, growth and environment: a sceptical review of 'The Skeptical Environmentalist' by B. Lomborg. *Scottish Journal of Political Economy* 51, 707–717.

Foster, S., Mommsen, W.J. and Robinson, R. (eds) (1988) *Bismarck, Europe and Africa: the Berlin Africa Conference 1884–1885 and the Onset of Partition*. Oxford University Press, Oxford, UK.

Fox, J. (2004) Swahili for the Broken-Hearted: Cape to Cairo (Book Review). *Getaway* (June), 147.

Freedom Park Trust (Undated) *Freedom Park: A Conceptual Framework* [brochure]. Pretoria, South Africa.

Frost, F. and Shanka, T. (2001) Cape to Cairo – can the dream be realised? *Journal of Vacation Marketing* 7, 235–244.

Gaborone, S. (1998) Land and resource tenure in Botswana: an overview. In: Rohoy, E. (ed.) *Proceedings of a Workshop on Land/Resource Tenure and Decentralisation*. Johannesburg, South Africa, pp. 53–65.

Gakwandi, A.S. (1996) Towards a political map of Africa. In: Abdul-Raheem, T. (ed.) *Pan Africanism: Politics, Economy and Social Change in the Twenty-first Century*. Pluto, London, UK, pp. 181–190.

Gale, W.D. (1935) *One Man's Vision*. Hutchinson, London, UK.

General Transfrontier Conservation and Resource Area Protocol (2000) *Protocol Between the Government of the Republic of South Africa, the Government of the Republic of Mozambique and the Government of the Kingdom of Swaziland*. Durban, South Africa.

Getaway (2002) Cape to Cairo (March), 126–127.

Getaway (2003) The wildest African dream (September), 61–68.

Getaway (2004) South Africa's newest national park (July), 43.

Gibb, R.A. (1996) Towards a new southern Africa: the challenge of regional economic co-operation and integration. *South African Journal of International Affairs* 4, 1–26.

Giliomee, H. (1992) 'Broedertwis': intra-Afrikaner conflicts in the transition from apartheid. *African Affairs* 91, 339–364.

Glassner, M. and Fahrer, C. (2004) *Political Geography* (3rd edn). Wiley, New York, USA.

Godwin, P. (2001) Without borders: uniting Africa's wildlife reserves. *National Geographic* (September), 6–29.

Grant, J.A. (1864) *A Walk Across Africa*. William Blackwood, Edinburgh, UK.

Grant, J.A. and Söderbaum, F. (eds) (2003) *The New Regionalism in Africa*. Ashgate, Aldershot, UK.

Green, J.E.S. (1936) *Rhodes Goes North*. G. Bell and Sons, London, UK.

Greenberg, S.B. (1987) Ideological struggles within the South African state. In: Marks, S. and Trapido, S. (eds) *The Politics of Race, Class and Nationalism in Twentieth Century South Africa*. Longman, London, UK, pp. 389–418.

Griffin, J., Cumming, D., Metcalfe, S., t'Sas-Rolfes, M., Singh, J., Chonguica, E., Rowen, M. and Oglethorpe, J. (1999) *Study on the Development of Transboundary Natural Resource Management Areas in Southern Africa*. Biodiversity Support Program, Washington, DC, USA.

Grogan, E.S. and Sharp, A.H. (1900) *From Cape to Cairo: The First Traverse of Africa from South to North*. Hurst and Blackett, London, UK.

Gross, F. (1956) *Rhodes of Africa*. Cassell and Company, London, UK.

Guardian (2006) South African tobacco tycoon who promoted equal rights during the apartheid era (23 January). London.

Gumede, M.V. (2005) Two-tier welcome. *BBC Focus on Africa* 16, 16–17.

Hagen, J. (2003) Redrawing the imagined map of Europe: the rise and fall of the 'centre'. *Political Geography* 22, 489–517.

Hall, M. (1907) *A Woman's Trek from the Cape to Cairo*. Methuen and Co., London, UK.

Hall-Martin, A. and Modise, S. (2002) *Status Report: Existing and Potential Transfrontier Conservation Areas in the SADC Region*. Stellenbosch, South Africa.

Hanks, J. (2000) Peace Parks: Categories, Principles and Criteria. Paper prepared for the IUCN meeting (Unpublished document). Gland, Switzerland.

Hanks, J. (2003) Transfrontier conservation areas (TFCAs) in southern Africa: their role in conserving biodiversity, socio-economic development and promoting a culture of peace. *Journal of Sustainable Forestry* 17, 127–148.

Harvey, D. (1996) *Justice, Nature and the Geography of Difference*. Blackwell, Cambridge MA, USA.

Hicks, R.W. (1991) Sehlabathebe National Park Conceptual Development Plan (Unpublished document). Department of Parks, Maseru, Lesotho.

Hirsch, A. (2005) *Season of Hope: Economic Reform under Mandela and Mbeki*. University of KwaZulu-Natal Press, Scottsville, South Africa.

Hoare, M. (1975) Kimberley and the Cape to Cairo Railway. Talk delivered to the Historical Society of Kimberley and the Northern Cape. Kimberley Archive, m5111.

Hobsbawm, E. (1994) *The Age of Empire: 1875–1914*. Abacus, London, UK.

Holgate, K. (2003) *Cape to Cairo: One Family's Adventures along the Waterways of Africa*. Struik, Cape Town, South Africa.

Hollander, J.M. (2003) *The Real Environmental Crisis: Why Poverty, not Affluence, is the Environment's Number One Enemy*. University of California Press, Berkeley CA, USA.

House of Commons (1894) *Treasury Minute, HE 3424 GB*. London, UK.

Hudson, D.J. and Murray, M.L. (1997) Possible effect of human population growth in the next twenty years on wildlife conservation. In: *Proceedings of a National Conference on Conservation and Management of Wildlife in Botswana*. Gaborone, Botswana, pp. 85–93.

Hughes, D.M. (2005) Third nature: making space and time in the Great Limpopo Conservation Area. *Cultural Anthropology* 20,157–184.

Hugon, A. (1993) *The Exploration of Africa: From Cairo to the Cape*. Thames and Hudson, London, UK.

Hulme, D. and Murphree, M. (1999) Communities, wildlife and the 'new conservation' in Africa. *Journal of International Development* 11, 277–285.

International Institute for Peace through Tourism (IIPT) (1999) Peace Park Dedication and Launch of Global Peace Park Project (11 November).

International Institute for Peace through Tourism (2003a) *Newsletter* (August).

International Institute for Peace through Tourism (2003b) Tanzania Action Plan. Dar es Salaam, Tanzania.

International Institute for Peace through Tourism (2005a) *Newsletter* (January).

International Institute for Peace through Tourism (2005b) Lusaka Declaration. Lusaka, Zambia.

International Tourism Monthly (2003) ECOCLUB Interview with Mr Frans Stroebel. http://ecoclub.com/news/051/interview.html (accessed on 9 Jan. 2006).

IUCN (2001) *Transboundary Protected Areas for Peace and Co-operation*. Washington, DC, USA.

Jhaveri, N. (2004) Petroimperialism: US oil interests and the Iraq War. *Antipode* 36, 2–11.

Jones, J.L. (2005) Transboundary conservation: development implications for communities in KwaZulu-Natal, South Africa. *International Journal of Sustainable Development and Ecology* 12, 266–278.

Jones, M. and MacLeod, G. (2004) Regional spaces, spaces of regionalism: territory, insurgent politics and the English question. *Transactions of the Institute of British Geographers* 29, 433–452.

Jones, T.B. (1995) *Wildlife Management, Utilisation and Tourism in Communal Areas. Research Discussion Paper No. 5*. Ministry of Environment and Tourism, Windhoek, Namibia.

KaNgwane (1993) *Annual Review*. Department of the Chief Minister. Kanyamazane, South Africa

Katerere, Y., Hill, R. and Moyo, S. (2001) *A Critique of Transboundary Natural Resource Management in Southern Africa*. IUCN, Harare, Zimbabwe.

Katzellenbogen, S. (1996) It didn't happen in Berlin: politics, economics and ignorance in the setting of Africa's colonial boundaries. In: Nugent, P. and Asiwaju, A.I. (eds) *African Boundaries: Barriers, Conduits and Opportunities*. Pinter, London, UK, pp. 21–34.

Kemp, A.C., Herholdt, J.J., White, I. and Harrison, J. (2001) Birds of the two largest national parks in South Africa: a method to generate estimates of population size for all species and assess their conservation ecology. *South African Journal of Science* 97, 393–403.

Kepe, T., Wynberg, R. and Ellis, W. (2005) Land reform and biodiversity conservation in South Africa: complimentary or in conflict? *International Journal of Biodiversity Science and Management* 1, 3–16.

Kgalagadi Transfrontier Agreement (1999) *Bilateral Agreement Between the Government of the Republic of Botswana and the Government of the Republic of South Africa on the Recognition of the Kgalagadi Transfrontier Park*. Pretoria, South Africa.

Kitchin, R. and Kneale, J. (2001) Science fiction or future fact? Exploring imaginative geographies of the new millennium. *Progress in Human Geography* 25, 19–35.

Kitschelt, H. (1986) Political opportunity structures and political protest: anti-nuclear Movements in four democracies. *British Journal of Political Science* 16, 57–58.

Krasner, S.D. (2004) Sharing sovereignty: new institutions for the collapsed and failing states. *International Security* 29, 85–120.

Kruger Mpumalanga International Airport (KMIA) (Undated) *KMI Airport* (Brochure). Nelspruit, South Africa.

Lamela, A., Molini, F. and Pérez, M. (2005) Geoistic planning: towards a worldwide territorial organisation. (Unpublished concept paper). Universidad Autonoma, Italy.

Langholz, J.A. and Lassoie, J. (2001) Perils and promise of privately owned protected areas. *BioScience* 51, 1079–1085.

Langlois, T.J. and Ballantine, W.J. (2005) Marine ecological research in New Zealand: developing predictive models through the study of no-take marine reserves. *Conservation Biology* 19, 1763–1770.

Lawrie, G. (1963) *New Light on South West Africa: Some Extracts from and Comments on the Odendaal Report*. Witwatersrand University Press, Johannesburg, South Africa.

Leabua, J. (1968) Speech by the Honourable Prime Minister Jonathan Leabua outlining [the] Development Policy for Lesotho and its Mountain Areas (1 December). National University of Lesotho Archives, Roma, Lesotho.

Le Billion, P. (2001) The political ecology of war: natural resources and armed conflicts. *Political Geography* 20, 561–584.

Lesotho (1990) Sehlabathebe National Park Management Plan (Unpublished document). Directorate of Oversea Surveys, Maseru, Lesotho.

Letcher, O. (1905) *The Bonds of Africa: Impressions of Travel and Sports from Cape Town to Cairo 1902–1912*. Long, London, UK.

Limpopo/Shashe Transfrontier Conservation Area (Undated) *Promotional Brochure*. Department of Environmental Affairs and Tourism, Pretoria, South Africa.

Lindegger, M. (2002) Permaculture for ecovillage design. In: Jackson, H. and Svensson, K. (eds) *Ecovillage Living: Restoring the Earth and her People*. Green Books, Dartington, UK, pp. 20–23.

Lomborg, B. (2001) *The Skeptical Environmentalist*. Cambridge University Press, Cambridge MA, USA.

Lubombo Spatial Development Initiative. (1998) [Pamphlets]. Department of Environmental Affairs and Tourism, South Africa.

Lunn, J. (1992) The political-economy of primary railway construction in the Rhodesias, 1890–1911. *Journal of African History* 33, 239–254.

Maathai, W. (2005) UCT Vice-Chancellor's Open Lecture (July). University of Cape Town, South Africa.

Mabogunje, A.L. (1984) The poor shall inherit the earth: issues of environmental quality and Third World development. *Geoforum* 15, 295–306.

MacArther, R.H. and Wilson, E.O. (1967) *The Theory of Island Biogeography*. Princeton University Press, Princeton NJ, USA.

MacLeod, F. (2004) Valli Moosa looks back. *Earthyear* 2, 7–11.

Madaune, C. (2002) Earth restoration in Colombia. In: Jackson, H. and Svensson, K. (eds) *Ecovillage Living: Restoring the Earth and her People*. Green Books, Dartington, UK, pp. 29–30.

Mail & Guardian (1999) The bittereinders of the Kalahari (July 2–8). Johannesburg, South Africa.

Mail & Guardian (2005) Game farming undermines reform (25 November – 1 December). Johannesburg, South Africa.

Mail & Guardian (2006) Rupert gatkruip makes one puke (27 January–2 February). Johannesburg, South Africa.

Malawi (2004) *Malawi National Development Strategy for Sustainable Development*. Ministry of Mines, National Resources and Environment, Lilongwe, Malawi.

Malcolm, D.O. (1939) *The British South Africa Company, incorporated by Royal Charter: 1889–1939*. British South Africa Company, London, UK.

Maloti Drakensberg Transfrontier Project (2006) *MDTP News* (March). Howick, South Africa.

Mamdani, M. (1996) *Citizen and Subject: Contemporary Africa and the Legacy of Late Colonialism*. Princeton University Press, Princeton NJ, USA.

Mandela, N.R. (1998) Speech at the launch of Lubombo Spatial Development Initiative (6 May). Durban, South Africa.

Mandela, N.R. (2001) Speech at a ceremony to celebrate the translocation of elephants from Kruger National Park to Mozambique (12 October). Kruger National Park, South Africa.

Marks, S. (1980) The myth of the empty land. *Africa Today* 30, 7–12.

Matthews, J.T. (1989) Redefining security. *Foreign Affairs* 68, 162–177.

Marston, S.A. and Smith, N. (2001) States, scales and households: limits to scale thinking? A response to Brenner. *Progress in Human Geography* 25, 615–619.

Maylam, P. (2005) *The Cult of Rhodes: Remembering an Imperialist in Africa*. David Philip, Cape Town, South Africa.

Mazrui, A.A. (1980) *The African Condition: A Political Diagnosis*. Heinemann, London, UK.

Mbaiwa, J.E. (2004) The success and sustainability of community-based natural resource management in the Okavango Delta, Botswana. *South African Geographical Journal* 86, 44–53.

Mbeki, G. (1984) *South Africa: The Peasant's Revolt*. International Defence and Aid Fund for Southern Africa, London, UK.

Mbeki, T. (2002) Speech on the signing of the Treaty on the Great Limpopo Transfrontier Park (9 December). Xai-Xai, Mozambique.

Mbeki, T. (2006) Apartheid is dead! Hail the spirit of Anton Rupert! *ANC Today* (10 February).

McCarthy, J. (2005) Scale, sovereignty and strategy in environmental governance. *Antipode* 37, 731–753.

McCormick, J. (1989) *The Global Environmental Movement: Reclaiming Paradise*. Belhaven Press, London, UK.

McGinnis, M.V. (1999) A rehearsal to bioregionalism. In: McGinnis, M.V. (ed.) *Bioregionalism*. Routledge, London, UK, pp. 1–9.

McKie, B.G., Pearson, R.G. and Cranston, P.S. (2005) Does bioregional history matter? Diversity and distribution of lotic midges (*Diptera: Chironomidae*) in the Australian Wet Tropics. *Austral Ecology* 30, 1–13.

Meadows, D.H., Meadows, D.L., Randers, J. and Berherens, W.W. (1972) *The Limits to Growth*. Universe Books, New York, USA.

Merrington, P. (2001) A staggered orientalism: the Cape-to-Cairo imaginary. *Poetics Today* 22, 333–364.

Metz, S. (2000) A strategic approach to African security: challenges and prospects. *African Security Review* 9, 3–18.

Michler, I. (2003) Bring down the fences. *Africa Geographic* 11, 78–84.

Miller, R. (1991) Postmaterialism and Green party activists in New Zealand. *Political Science* 43, 43–66.

Mills, G. and Streamlau, J. (1999) *The Privatisation of Security in Africa*. Institute of International Affairs, Johannesburg, South Africa.

Mirror (2004) Regional development through conservation (10 October). Makhado, South Africa.

Mittermeier, R.A., Myers, N., Thomsen, J.B. and da Fonscca, G.A.B. (1998) Biodiversity hotspots and major wilderness areas: approaches to setting conservation priorities. *Conservation Biology* 12, 516–520.

Mkondiwa, G. (2003) Opening Speech to the Meeting of Donors and Partners for Lake Malawi/Niassa/Nyasa Ecoregion (12–13 May). Club Makokola, Malawi.

Mongwe, A. and Tevera, D. (2000) Land rights and Basarwa people in Botswana. In: Tevera, D. and Moyo, S. (eds) *Environmental Security in Southern Africa*. SAPES Books, Harare, Zimbabwe, pp. 75–88.

Moore, P. (2003) *Swahili for the Broken-Hearted: Cape Town to Cairo by Any Means Possible.* Bantam, London, UK.

Moosa, V. (2000) Address at the launch of the Kgalagadi Transfrontier Park (12 May). Gaborone, Botswana.

Mswati, (King) (1998) Speech at the launch of Lubombo Spatial Development Initiative (6 May). Durban, South Africa.

Mudimbe, V.Y. (1988) *The Idea of Africa*. Indiana University Press, Bloomington, IN, USA.

Müller-Rommel, F. (2002) The lifespan and the political performance of green parties in Western Europe. *Environmental Politics* 11, 1–16.

Murray, R. (1996) *Mister Africa: The Story of African Car Hire and Trans Africa Safaris and the Development of Overland Touring from Cape Town to the Nile.* Murray, Cape Town, South Africa.

Muthien, Y. and Khosa, M. (1998) *Regionalism in the New South Africa*. Ashgate, Aldershot, UK.

Nabudere, D.W. (2001) African unity in historical perspective. In: Maloka, E. (ed.) *A United States of Africa*. Africa Institute of South Africa, Pretoria, South Africa, pp. 9–28.

Namibia. (1998) *Human Development Report*. Windhoek, Namibia.

Nature Conservancy (Undated) *The Power of Partnerships*. Arlington, VA, USA.

Nature Conservancy (2001) *Private Land Management and Conservation: A Guide to Organisations*. Arlington, VA, USA.

Nelson, J. (1996) From Cape to Cairo: stressing Africa's geographical heritage. *Southern Social Studies Journal* 21, 41–50.

Neumann, R.P (2004) Moral and discursive geographies in the war for biodiversity in Africa. *Political Geography* 23, 813–837.

Nevins, J. (2002) *Operation Gatekeeper: The Rise of the 'Illegal Alien' and the Making of the US–Mexico Boundary*. Routledge, New York, USA.

New Ecologist (1979) Ecology Party Manifesto (March/April), 59–61.

Newman, D. and Paasi, A. (1998) Fences and neighbours in the postmodern world: boundary narratives in political geography. *Progress in Human Geography* 22, 186–207.

Newmark, W.D. (ed.) (1991) *The Conservation of Mount Kilimanjaro*. IUCN, Gland, Switzerland.

New Partnership for Africa's Development (NEPAD) (2001) *The New Partnership for Africa's Development* (October). Midrand, South Africa.

New Partnership for Africa's Development (2002a) *Towards the Implementation of the New Partnership for Africa's Development: Progress Report and Initial Action Plan* (June). Midrand, South Africa.

New Partnership for Africa's Development (2002b) Main Report (Draft): Short-term Action Plan Infrastructure (May). Midrand, South Africa.

New Partnership for Africa's Development (2003) *Action Plan for the Environment Initiative*. Midrand, South Africa.

Nkonoki, S.R. (1983) *The East African Community in Retrospect: The Context of Regional Co-operation in Eastern Africa. DERAP Working Paper A 287*. Bergen, Norway.

Nkrumah, K. (1961) *I Speak of Freedom*. Panaf Books, London, UK.

North, P. (2005) Scaling alternative economic practices? Some lessons from alternative currencies. *Transactions of the Institute of British Geographers* 30, 221–233.

Nugent, P. and Asiwaju, A.I. (eds) (1996) *African Boundaries: Barriers, Conduits and Opportunities*. Pinter, London, UK.

Nyankale, H.S. (1979) *Evaluation of Sehlabathebe National Park* (Research Project). National University of Lesotho, Roma, Lesotho.

Nyasulu, K.M. (1997) SADC FSTCU SADC Focal Point. *Natural Newsletter*. Lilongwe, Malawi.

Oberholzer, O. (1996) *Beyond Bagamoyo*. Hotazel, Cape Town, South Africa.

Omo-Fadaka, J. (1972) 'Tanzanian Way' to self reliance. *Ecologist* 2, 7–11.

O'Riordan, T. (1996) Environmentalism on the move. In: Douglas, I., Huggett, R. and Robinson, M. (eds) *Companion Encyclopaedia of Geography: The Environment and Humankind*. Routledge, London, UK, pp. 449–476.

O'Tuathail, G. (1996) *Critical Geopolitics*. University of Minnesota Press, Minneapolis, MN, USA.

O'Tuathail, G. (1998) Postmodern geopolitics? The modern geopolitical imagination and beyond. In: O'Tuathail, G. and Dalby, S. (eds) *Rethinking Geopolitics*. Routledge, London, UK, pp. 16–38.

Paasi, A. (1996) *Territories, Boundaries and Consciousness: The Changing Geographies of the Finish-Russian Border*. Wiley, London, UK, pp. 449–476.

Paasi, A. (2002) Bounded spaces in the mobile world: deconstructing 'regional identity'. *Tijdschrift Voor Economische en Sociale Geografie* 93, 137–148.

Paasi, A. (Forthcoming) Generations and the 'development' of border studies. *Geopolitics*.

Painter, J. (1995) *Politics, Geography and 'Political Geography': A Critical Perspective*. Arnold, London, UK.

Peace Parks Foundation (PPF) (1997) *Annual Report*. Stellenbosch, South Africa.

Peace Parks Foundation (1998) *Annual Report*. Stellenbosch, South Africa.

Peace Parks Foundation (1999) New university brings hope for global peace and security. *Newsletter* (3rd edn), Stellenbosch, South Africa.

Peace Parks Foundation (2000) *Annual Report*. Stellenbosch.

Peace Parks Foundation (2002) The facilitation and development of transfrontier conservation areas in southern Africa: funding application in support of Malolotja/Songimvelo TFCA Co-ordinator (Unpublished document). Stellenbosch, South Africa.

Peace Parks Foundation (2003) *US Congress support for peace parks*. Media Release (15 July). Stellenbosch, South Africa.

Pedynowski, D. (2003) Toward a more reflexive enviromentalism; ecological knowledge and advocacy in the crown of the continent ecosystem. *Society and Natural Resources* 16, 807–825.

Pinheiro, I., Gabaake, G. and Heyns, P. (2003) Co-operation in the Okavango river basin: The OKACOM perspective. In: Turton, A., Ashton, P. and Cloete, E. (eds) *Transboundary Rivers, Sovereignty and Development: Hydropolitical Drivers in the Okavango River Basin*. African Water Issues Research Unit and Green Cross International, Pretoria (South Africa) and Geneva, Switzerland, pp. 105–118.

Pinnock, D. (1996) Superparks: the impossible dream? *Getaway* (November), 88–97.

Player, I. (1997) *Zululand Wilderness: Shadow and Soul*. David Philip, Cape Town, South Africa.

Pollard, J.S. and Sidaway, J.D. (2002) Editorial: Euroland-economic, cultural and political geographies. *Transactions of the Institute of British Geography* 27, 7–10.

Porritt, J. (1979) Gearing up for the General Election. *New Ecologist* (Jan/Feb), 20–22.

Pulido, L. (1996) A critical review of the methodology of environmental racism research. *Antipode* 28, 142–159.

Purcell, M. and Nevins, J. (2005) Pushing the boundary: state restructuring, state theory, and the case of US–Mexico border enforcement in the 1990s. *Political Geography* 24, 211–235.

Rainer, H., Asuma, S., Gray, M., Kalpers, J., Kayitare, A., Rutagarama, E., Sivha, M. and Lanjouw, A. (2003) Regional conservation in the Virunga–Bwindi Region: the impact of transfrontier collaboration through the experiences of the Internal Gorilla Conservation Programme. *Journal of Sustainable Forestry* 17, 189–204.

Ramutsindela, M. (1998) Afrikaner nationalism, electioneering, and the politics of the volkstaat. *Politics* 18, 179–188.

Ramutsindela, M. (2002) The perfect way to ending a painful past? Makuleke land deal in South Africa. *Geoforum* 33, 15–24.

Ramutsindela, M. (2004a) Glocalisation and nature conservation strategies in 21st century southern Africa. *Tijdscrift voor Economische en Sociale Geografie* 95, 61–72.

Ramutsindela, M. (2004b) *Parks and People in Postcolonial Societies: Experiences in Southern Africa*. Kluwer, Dordrecht, The Netherlands.

Ramutsindela, M.F. and Simon, D. (1999) The politics of territory and place in post-apartheid South Africa: the disputed area of Bushbuckridge. *Journal of Southern African Studies* 25, 479–498.

Ranger, T. (1999) *Voices from the Rocks: Nature, Culture and History in the Matopos Hills of Zimbabwe*. James Currey, Oxford, UK.

Raphael, L.A.C. (1936) *The Cape to Cairo Dream: A study in British Imperialism*. Columbia University Press, New York, USA.

Redfern, D. (1999) *Cape to Cairo – Solo*. Connecting: Solo Travel Network, Canada.

Regional Tourism Organisation of Southern Africa (RETOSA) (1998) *Marketing Strategy Plan, 1998–2002*. Halfway House, South Africa.

Regional Tourism Organisation of Southern Africa (1999) *Southern Africa Tourism News*. Halfway House, South Africa.

Robertson, D. (1999) Rail deal links Cape to Cairo. *Business Times* (online).

Rodrigues, M. (2002) Cape to Cairo tourism dream unfolds. *Engineering News Online.*

Rozemeijer, N. and Van der Jagt, C. (2000) *Community-based Natural Resource Management in Botswana: How Community-based is Community Natural Resource Management in Botswana, Occasional Paper Series*. IUCN/SNV CBNRM Support Programme, Gaborone, Botswana.

Rudig, W. and Lowe, P.D. (1986) The withered greening of British politics: a study of the Ecology Party. *Political Studies* 34, 262–284.

Ryan, B. (1999) From war zone to peace parks: a forgotten corner of Africa. *Optima* 45, 52–64.

Said, E. (1994) *Culture and Imperialism*. Vintage, London, UK.

Sale, K. (1985) *Dwellers in the Land: The Bioregional Vision*. Sierra Club Books, San Francisco CA, USA.

Sandwith, T. (2003) Overcoming barriers: conservation and development in the Maloti–Drakensberg Mountains of southern Africa. *Journal of Sustainable Forestry* 17, 149–170.

Sandwith, T., Shine, C., Hamilton, L. and Sheppard, D. (2001) *Transboundary Protected Areas for Peace and Co-operation*. IUCN, Gland, Switzerland.

Savage, T. and Rahall, J. (2003) Sierra Leone: history hidden by horror. In: Doxtader, E. and Villa-Vincencio, C. (eds) *Through Fire with Water: The Roots of Division and the Potential for Reconciliation in Africa*. Institute for Justice and Reconciliation, Cape Town, South Africa.

Sayre, N.F. (2005) Ecological and geographical scale: parallels and potential for integration. *Progress in Human Geography* 29, 276–290.

Secretariat of the Convention on Biological Diversity (2001) *Handbook of the Convention on Biological Diversity*. Montreal, Canada.

Shalom, S.R. (1999) The continuity of U.S. imperialism. *New Politics*, 26.

Shechambo, F., Salebe, J. and Mariki, S. (2001) *Report on the National Awareness Workshop on the Economic Value of Tanzania's Forest Ecosystems* (15 August). Dar es Salaam, Tanzania.

Sheppard, D. (1999) Conservation without frontiers – the global view. Paper presented to EUROPARC 99 Conference (15–19 September), Zakopane, Poland.

Shumba, M. (1998) Devolving of natural resource tenure in Namibia. In: Rehoy, E. (ed.) *Natural Resources Tenure in Southern Africa: Exploring Options and Opportunities – Proceedings of a Workshop on Land/Resource Tenure and Decentralisation*. Johannesburg, South Africa, pp. 74–82.

Sidaway, J.D. (2002) *Imagined Regional Communities*. Routledge, London, UK.

Simon, D. (1998) Introduction: shedding the past, shaping the future. In: Simon, D. (ed.) *South Africa in Southern Africa: Reconfiguring the Region*. James Currey, Oxford, UK, pp. 1–16.

Simon, D. (2003) Regional development-environment discourses, policies and practices in post-apartheid southern Africa. In: Grant, J.A. and Söderbaum, F. (eds) *New Regionalisms in Africa*. Ashgate, Aldershot, UK, pp. 67–89.

Smith, P. (2005) Mercenaries: dog days for Africa's hired guns. *Africa Report* (May), 22–24.

South Africa (2000) Gaza–Kruger–Gonarezhou Transfrontier Park Agreement. Department of Environmental Affairs and Tourism (Unpublished document). Pretoria, South Africa.

South Africa (2003) Parks that transcend borders (11 August). *South Africa Government Online*.

South African Airlink (2003) *The Arch in the Kruger* [In-Flight Magazine] (October). Johannesburg, South Africa.

South African National Parks (SANParks) (1997) *News Clip*. Pretoria, South Africa.

South African National Parks (2004) *Annual Report*. Pretoria, South Africa.

Southern African Development Community (SADC) (1992) *Treaty of the Southern African Development Community*. Gaborone, Botswana.

Southern African Development Community (1993) *Summit Communique* (5 September). Mbabane, Swaziland.

Southern African Development Community (1994) *Summit Communique* (29 August). Gaborone, Botswana.

Southern African Development Community (1995a) *Protocol on Shared Watercourse Systems*. Gaborone, Botswana.

Southern African Development Community (1995b) *Summit Communique* (28 August). Johannesburg, South Africa.

Southern African Development Community (1996a) *Caring for Southern Africa's Resources. Environment and Land Management Sector*. Maseru, Lesotho.

Southern African Development Community (1996b) *SADC Policy and Strategy for Environment and Sustainable Development: Toward Equity-led Growth and Sustainable Development in Southern Africa*. Environment and Land Management Sector Co-ordination Unit, Maseru, Lesotho.

Southern African Development Community (1997) *Review and Rationalisation of the SADC Programme of Action: Main Report (Volume 2)*. Gaborone, Botswana.

Southern African Development Community (1999a) *Summit* (18 August). Maputo, Mozambique.

Southern African Development Community (1999b) *Protocol on Wildlife Conservation and Law Enforcement*. Lilongwe, Malawi.

Southern African Development Community (2002) *Biodiversity Focal Point Progress Report, April 2000–December 2001*. Lilongwe, Malawi.

Southern African Development Community (2005) *Official SADC Trade, Industry and Investment Review: 25th Anniversary Commemorative Edition*. Gaborone, Botswana.

Southern African Development Co-ordination Conference (SADCC) (1980) *Summit Communique* (1 April). Lusaka, Zambia.

Southern African Development Co-ordination Conference (1981) *Summit Communique* (20 July). Salisbury, Zimbabwe.

Southern African Development Co-ordination Conference (1983) *Summit Communique* (11 July). Maputo, Mozambique.

Southern African Development Co-ordination Conference (1984) *Summit Communique* (6 July). Gaborone, Botswana.

Southern African Development Co-ordination Conference (1989) *Summit Communique* (25 August). Harare, Zimbabwe.

Southern African Development Co-ordination Conference (1991) *Summit Communique* (26 August). Arusha, Tanzania.

Southern African Nature Foundation (SANF) (1984) *Annual Report*. Stellenbosch, South Africa.

Southern African Nature Foundation (1985) *Annual Report*. Stellenbosch, South Africa.

Southern African Nature Foundation (1986) *Annual Report*. Stellenbosch, South Africa.

Southern African Nature Foundation (1987) *Annual Report*. Stellenbosch, South Africa.

Southern African Nature Foundation (1988) *Annual Report*. Stellenbosch, South Africa.

Southern African Nature Foundation (1989) *Annual Report*. Stellenbosch, South Africa.

Southern African Nature Foundation (1990) *Annual Report*. Stellenbosch, South Africa.

Southern African Nature Foundation (1991) *Annual Report*. Stellenbosch, South Africa.

Southern African Nature Foundation (1993) *25th Anniversary Review*. Stellenbosch, South Africa.

Sparks, A. (2003) *Tomorrow is Another Country*. Jonathan Ball, Johannesburg, South Africa.

Spenceley, A. (Forthcoming) Peace parks as ecological systems: testing environmental resilience in southern Africa. In: Ali, S.A. (ed.) *Peace Parks: Conservation and Conflict Resolution*. MIT Press, Cambridge MA, USA.

Spretnak, C. and Capra, A.F. (1984) *Green Politics: The Global Promise*. Hutchinson, London, UK.

Star (2006a) Mooi loop, Anton (20 January). Johannesburg, South Africa.

Star (2006b) Rupert will be remembered for his values (20 January). Johannesburg, South Africa.

Star (2006c) Rupert praised as 'man who saw miracles' (26 January). Johannesburg, South Africa.

Strage, M. (1973) *Cape to Cairo*. Jonathan Cape, London, UK.

Sub-Regional Environmental Action Plan (2006) Draft Document of the Southern Africa Sub-Regional Environmental Action Plan of the Environmental Action Plan of NEPAD (Unpublished document). Pretoria, South Africa.

Sumaye, F. (2003) Speech by Prime Minister of the United Republic of Tanzania (8 December). Dar es Salaam, Tanzania.

Sunday Independent (2006a) There's a disappearing dearth of dirt on the barefoot billionaire (22 January). Johannesburg, South Africa.

Sunday Independent (2006b) Pragmatist who advised 'one man, one job' (22 January). Johannesburg, South Africa.

Sunday Times (2006a) The benevolent tycoon (22 January). Johannesburg, South Africa.

Sunday Times (2006b) Careers (21 May). Johannesburg, South Africa.

Svensson, K. (2002) What is an ecovillage? In: Jackson, H. and Svensson, K. (eds) *Ecovillage Living: Restoring the Earth and her People*. Green Books, Dartington, UK, pp. 10–12.

Tabor, G. (2003) *The Cape to Cairo Railway and River Routes*. Genta, London, UK.

Tanner, R., Freimund, W., Hayden, B. and Dolan, B. (Forthcoming) The Waterton-Glacier International Peace Park: conservation amidst border security. In: Ali, S.A. (ed.) *Peace Parks: Conservation and Conflict Resolution*. MIT Press, Cambridge MA, USA.

Taylor, I. and Mokhawa, G. (2003) Not forever: Botswana, conflict diamonds and the bushmen. *African Affairs* 102, 261–283.

Tema, W. (2000) A press release from the Kalahari Conservation Society on the occasion to the launching of the Kgalagadi Transfrontier Park (Unpublished document). Gaborone, Botswana.

Theroux, P. (2003) *Dark Star Safari: Overland from Cairo to Cape*. Penguin, London, UK.

Thrift, N. (1999) Steps to an ecology of place. In: Massey, D., Allen, J. and Sarre, P. (eds) (1999) *Human Geography Today*. Polity, Cambridge, UK, pp. 295–322.

Tiley, S. (2004) *Mapungubwe: South Africa's Crown Jewels*. Sunbird, Cape Town, South Africa.

Town and Regional Planning Commission (2000) *An Assessment of the Need for, and the Role of, a Regional Authority for the Drakensberg*. Pietermaritzburg, South Africa.

Town and Regional Planning Commission (2001) *A Special Case Area Plan for the Drakensberg*. Pietermaritzburg, South Africa.

Treatt, S.C. (1927) *Cape to Cairo: The Record of a Historic Motor Journey*. George Harrap and Co, London, UK.

Treaty of Ais-Ais/Richtersveld (2003) Treaty between the Government of the Republic of South Africa and the Government of the Republic of Namibia on the Establishment of the Ais-Ais/Richtersveld Transfrontier Park. Pretoria, South Africa

Treaty of the Great Limpopo Transfrontier Park (2002) Treaty between the Government of the Republic of Mozambique, the Government of the Republic of South Africa and the Government of the Republic of Zimbabwe on the Establishment of the Great Limpopo Transfrontier Park. Pretoria, South Africa.

Trilateral Memorandum of Understanding (2004) Central Albertine Rift Transfrontier Protected Area Network (Unpublished document).

Tukahirwa, E. (2002) Introducing the African Protected Areas Initiative (APAI). Paper presented to the IUCN-WSSD Preparatory Meeting (23–25 April). Dakar, Senegal.

Turnbull, M. (2001) Breaking boundaries. *Africa Environment & Wildlife* (May), 60–70.

Turnbull, M. (2002) Conserving abundance. *Africa Geographic*, 65–73.

Turton, A.R. (2003) The hydropolitical dynamics of co-operation in Southern Africa. In: *Transboundary rivers, sovereignty and development: Hydropolitical drivers in the Okavango River basin*. African Water Issues Research Unit (AWIRU) and Green Cross International (GCI), pp. 83–103.

Twyman, C. (1998) Rethinking community resource management: managing resources or managing people in Botswana? *Third World Quarterly* 19, 745–770.

United Nations Educational, Scientific and Cultural Organisation (UNESCO) (2000) World Heritage Committee Inscribes 61 New Sites on World Heritage List. *News* (30 November). Paris, France.

United Nations Environment Programme (UNEP) (2002) *Africa Environment Outlook*. Nairobi, Kenya.

Van Aarde, R. and Jackson, T. (2006) Elephants in Africa. *Africa Geographic* (April), 28–29.

Van Ameron, M. and Büscher, B. (2005) Peace parks in southern Africa: bringers of an African renaissance? *Journal of Modern African Studies* 43, 159–182.

Van Vuuren, I. (1998) The changing nature of warfare: implications for Africa. *African Security Review* 7, 54–65.

Visvanathan, S. (1998) Mrs Brundland's disenchanted cosmos. In: O'Tuathail, G., Dalby. S. and Routledge, P. (eds) *The Geopolitical Reader*. Routledge, London, UK, pp. 231–235.

Vreugdenhill, D., Terborgh, J., Cleef, A.M., Sinistsym, M., Boere, G.C., Archaga, V.L. and Prins, H.H.T. (2003) *Comprehensive Protected Areas System Composition and Monitoring*. World Institute for Conservation and Environment, Shepherdstown WV, USA.

Waites, B. (1984) The search for Britain's regions. *Ecologist* 14, 161–164.

Waters, W. (2002) Mapping Schengeland: denaturalizing the border. *Environment and Planning D: Society and Space* 20, 561–580.

Weinthal, L. (1922) *The Story of Cape to Cairo Railway and River Route* (vol. 1). Pioneer Publishing, London, UK.

Westing, A.H. (1998) Establishment and management of transfrontier reserves for conflict prevention and confidence building. *Environmental Conservation* 25, 91–94.

WildNet Africa (1999) Kgalagadi Transfrontier Park: a first for southern Africa. www.wildnetafrica.com (accessed 20/02/2001).

Williams, R. (1922) The Cape to Cairo railway from the point of view of African development. *Journal of the Central Asian Society* 9, 147–166.

Williams, R. (1998) Reallocating Defence Expenditures for development: the South African experience. *African Security Review* 7, 33–47.

Wills, J.T. (Undated) 'The Cape to Cairo' The Bulawayo–Tanganyika and other railways. Pamphlets on the South African War [Unpublished]. Institute of Commonwealth Studies, London, UK.

Wits Refugee Research Programme (2002) *A Park for the People? Great Limpopo Transfrontier Park – Community Consultation in Coutanda 16, Mozambique.* Preliminary Report (March). Acornhoek, South Africa.

Wolfensohn, J., Seligmann, P. and El-Ashry, M. (2000) Winning the war on biodiversity conservation. *New Perspective Quarterly* 17, 38–39.

Wolmer, W. (2003) Transboundary conservation: the politics of ecological integrity in the Great Limpopo Transfrontier Park. *Journal of Southern African Studies* 29, 261–278.

World Bank (1996) *Mozambique: Transfrontier Conservation Areas Pilot and Institutional Strengthening Project, Report No. 15543–MOZ.* Washington, DC, USA.

Yeld, J. (2006) The passing of a legend: Anton Rupert. *African Wildlife* 60, 31

Young, Z. (2002) *A New Green Order? The World Bank and the Politics of the Global Environmental Facility.* Pluto, London, UK.

Zbicz, D.C. and Green, M.J.B. (1997) Status of the world's transfrontier protected areas. *Parks* 7, 5–10.

Zoutpansberger (2004) Mapungubwe officially opened (1 October). Makhado, South Africa.

Index

O

P

Printed and bound by CPI Group (UK) Ltd, Croydon, CR0 4YY

16/09/2026

14963056-0004